ORGANIC CHEMISTRY 2 PRACTICE PROBLEMS 2014

HAROLD M. ICKES II, M.D.
AND
ARTHUR MARTIN, PH.D.

COVER PHOTO: Courtesy of William C. Dennis, Jr.

Special thanks to Prof. Rhett C. Smith of Clemson University for providing us with problems for use in this text.

Ideal Quill Publishing Group
idealquill.com
For information or to report typographical errors and suggestions for improvement, contact our editorial office: info@idealquill.com

Printed in the United States of America

10 9 8 7 6 5 4 3 2 1

ISBN 978-0615943206

Organic Chemistry 2

Practice Problems 2014

by Harold M. Ickes II, M.D. and

Arthur Martin, Ph.D.

Table of Contents

Problem Set 1: Alcohols 5

Problem Set 2: Epoxides and other Ethers 21

Problem Set 3: Organometallic Reagents and Multistep Synthesis 38

Problem Set 4: Aromaticity, Benzene and Electrophilic Aromatic Substitution on Benzene 60

Problem Set 5: Electrophilic Aromatic Substitution on Substituted Benzene 78

Problem Set 6: More Reactions of Arenes: Beyond Electrophilic Aromatic Substitution 95

Problem Set 7: Nucleophilic Acyl Substitution 109

Problem Set 8: Nucleophilic Addition on Carbonyls 119

Problem Set 9: Reactions involving the Carbonyl ⍰-Carbon 138

Problem Set 10: Reactions of Carbonyls with Amines / Amine Derivatives; Reactions of Nitriles and Imines 160

Progress Check 1: 168

Key for Progress Check 1: 177

Detailed Feedback for Progress Check 1 178

Progress Check 2: 200

Key for Progress Check 2: 208

Detailed Feedback for Progress Check 2 209

Progress Check 3: 235

Key for Progress Check 3: 243

Detailed Feedback for Progress Check 3 244

Problem Set 1: Alcohols

Date(s) covered in class:

Date by which material should be studied:

Chapter(s) in text covering material:

Date this problem set is due:

Points this problem set is worth:

1. Rank the following in terms of boiling point, from lowest to highest. Explain your rankings in terms of intermolecular forces prevalent in a sample of each molecule.

(A)

(B)

(C)

(D)

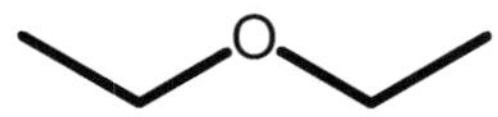

2. Provide the major product(s) of the following reaction, including stereochemistry where relevant. Show a reasonable arrow-pushing mechanism by which the major product could form. Label intermediates and transition states where they occur in your mechanism. For concerted reactions, always draw the transition state.

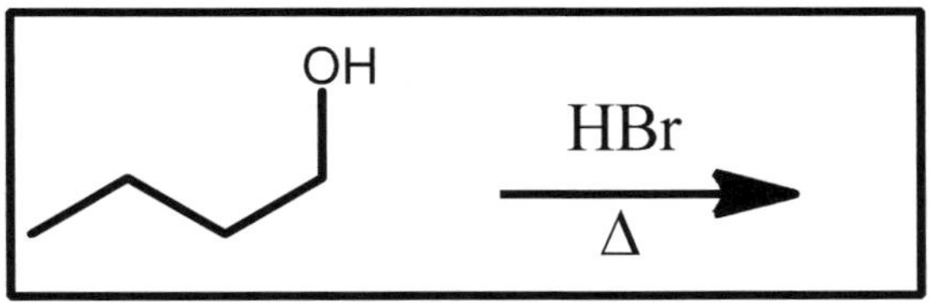

3. Provide the major product(s) of the following reaction, including stereochemistry where relevant. Show a reasonable arrow-pushing mechanism by which the major product could form. Label intermediates and transition states where they occur in your mechanism. For concerted reactions, always draw the transition state.

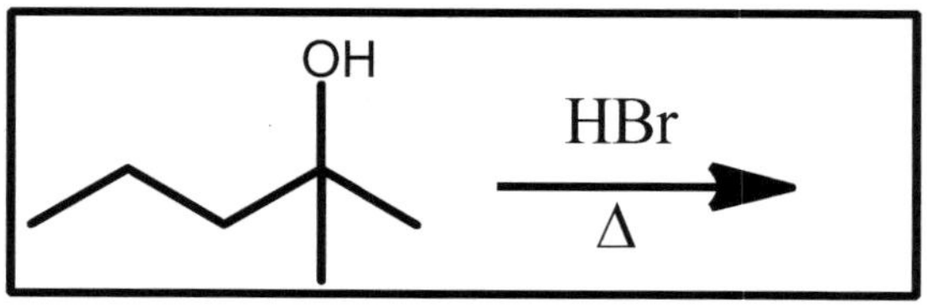

4. Provide the major product(s) of the following reaction, including stereochemistry where relevant. Show a reasonable arrow-pushing mechanism by which the major product could form. Label intermediates and transition states where they occur in your mechanism. For concerted reactions, always draw the transition state.

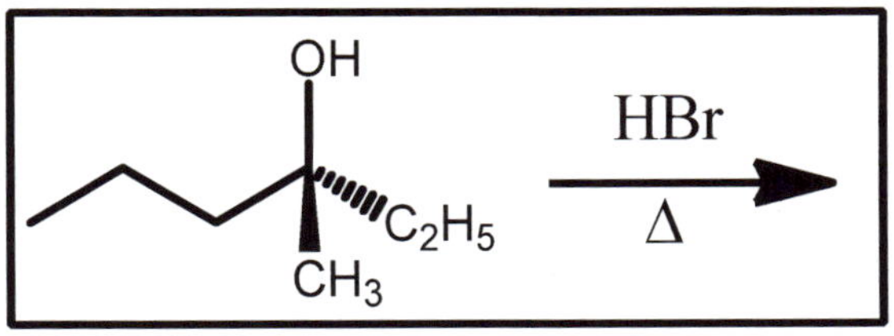

5. Provide the major product(s) of the following reaction, including stereochemistry where relevant. Show a reasonable arrow-pushing mechanism by which the major product could form. Label intermediates and transition states where they occur in your mechanism. For concerted reactions, always draw the transition state. Be sure to draw the structure of pyridine and to indicate its role in the reaction.

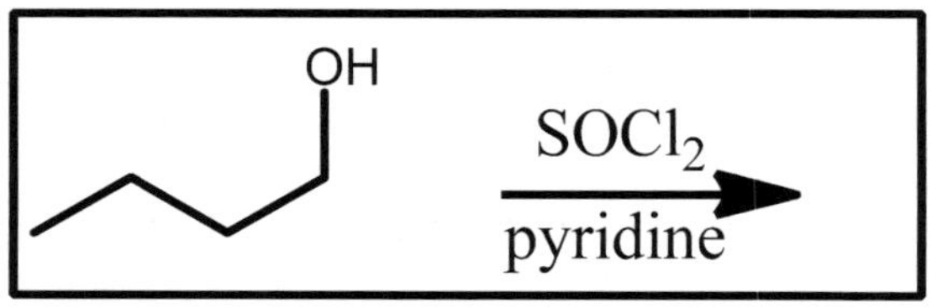

6. Provide the major product(s) of the following reaction, including stereochemistry where relevant. Show a reasonable arrow-pushing mechanism by which the major product could form. Label intermediates and transition states where they occur in your mechanism. For concerted reactions, always draw the transition state.

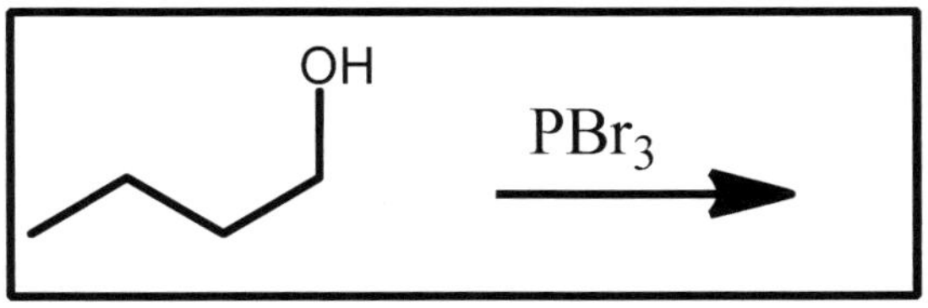

7. Provide the major product(s) of the following reaction, including stereochemistry where relevant. Show a reasonable arrow-pushing mechanism by which the major product could form. Label intermediates and transition states where they occur in your mechanism. For concerted reactions, always draw the transition state.

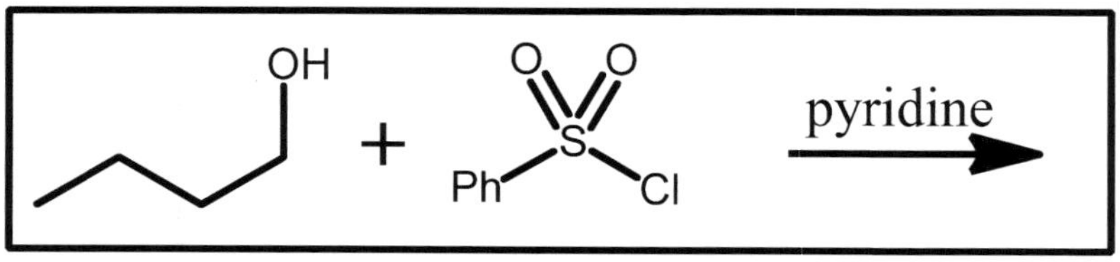

8. Provide the structures and names for the anions that are abbreviated $^{-}$OTs and $^{-}$OTf.

9. Provide the major product(s) of the following reaction, including stereochemistry where relevant. Show a reasonable arrow-pushing mechanism by which the major product could form. Label intermediates and transition states where they occur in your mechanism. For concerted reactions, always draw the transition state.

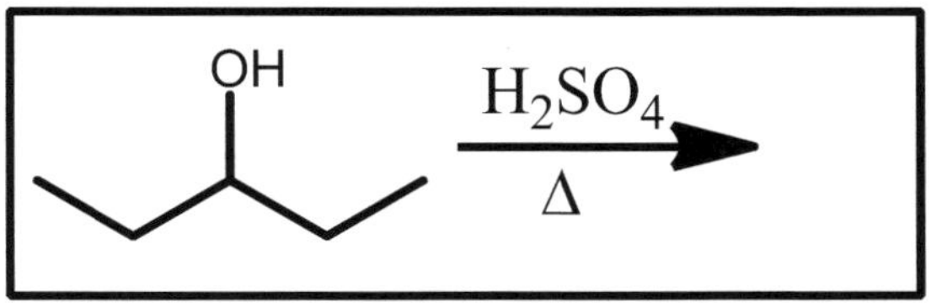

10. Provide the major product(s) of the following reaction, including stereochemistry where relevant. Show a reasonable arrow-pushing mechanism by which the major product could form. Label intermediates and transition states where they occur in your mechanism. For concerted reactions, always draw the transition state.

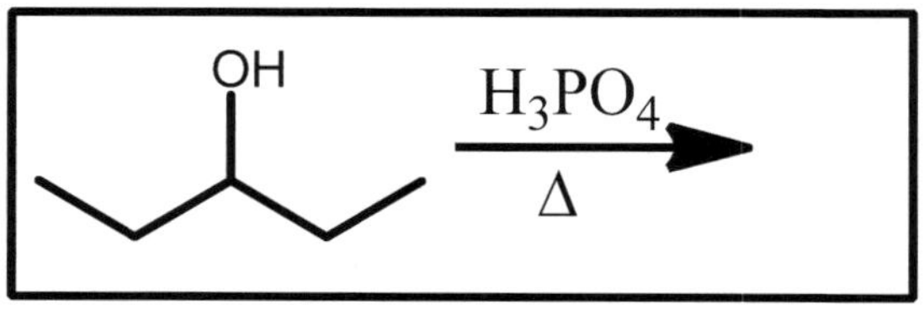

11. Provide a statement of Zaitsev's Rule.

12. Provide the major product(s) of the following reaction, including stereochemistry where relevant. Show a reasonable arrow-pushing mechanism by which the major product could form. Label intermediates and transition states where they occur in your mechanism. For concerted reactions, always draw the transition state.

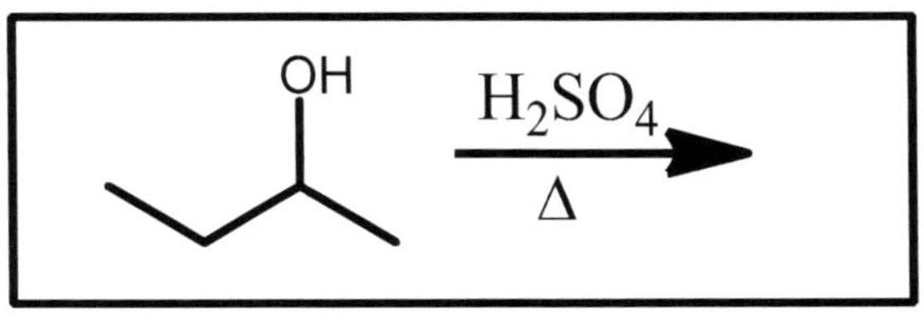

13a. Give an example of a 1,2-hydride shift:

13b. Give an example of a 1,2-methyl shift:

13c. What is the driving force for these carbocation rearrangements?

14. Provide the major product(s) of the following reaction, including stereochemistry where relevant. Show a reasonable arrow-pushing mechanism by which the major product could form. Label intermediates and transition states where they occur in your mechanism. For concerted reactions, always draw the transition state.

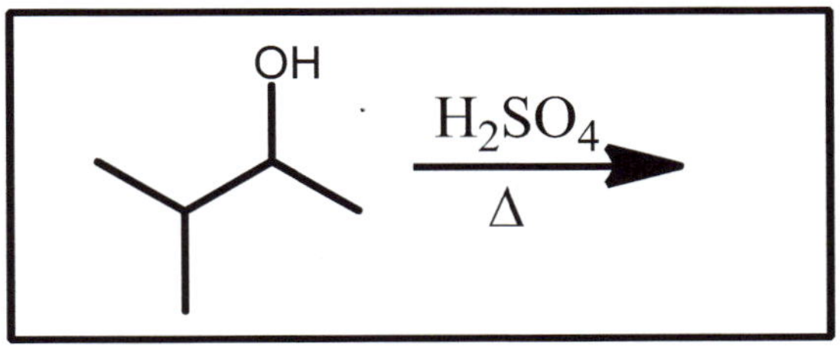

15. Provide the major product(s) of the following reaction, including stereochemistry where relevant. Show a reasonable arrow-pushing mechanism by which the major product could form. Label intermediates and transition states where they occur in your mechanism. For concerted reactions, always draw the transition state.

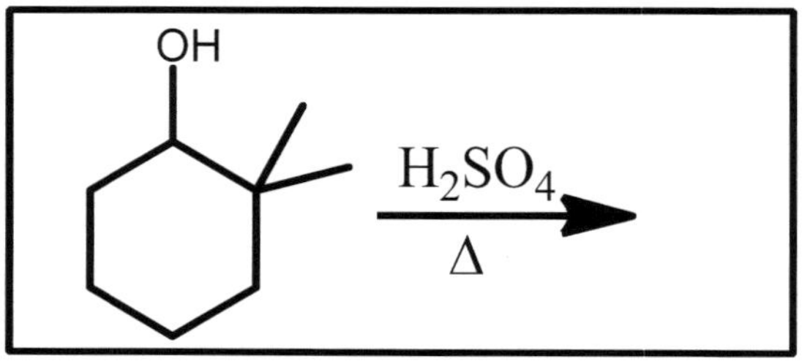

16. Rank the molecules whose structures are provided in terms of the rate (fastest to slowest) at which they would undergo elimination upon heating in sulfuric acid. Explain your reasoning.

(A)

(B)

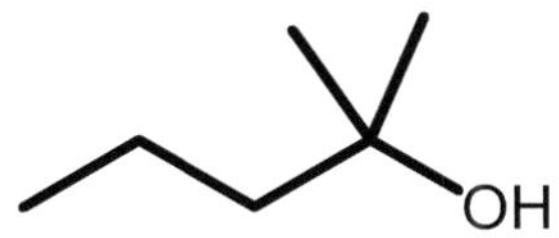

(C)

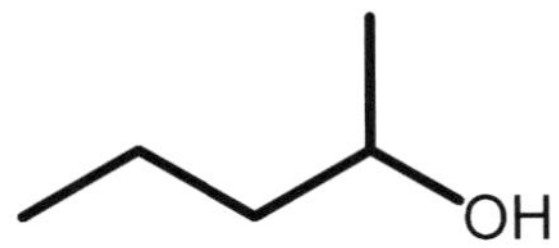

17. Provide the major products of the following reactions:

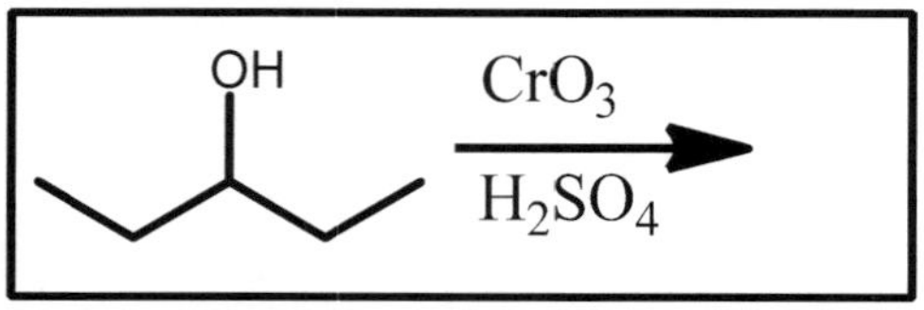

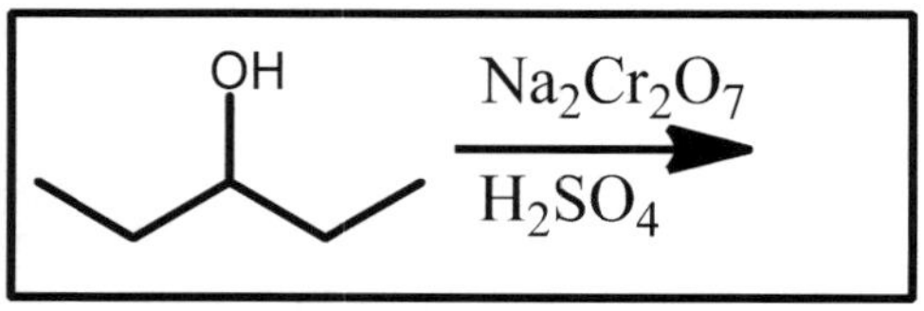

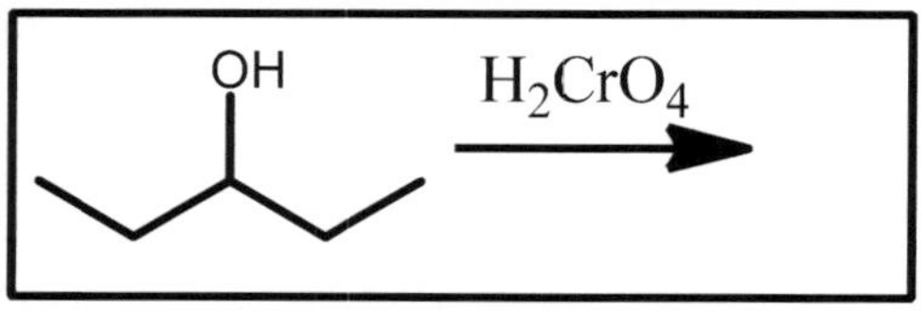

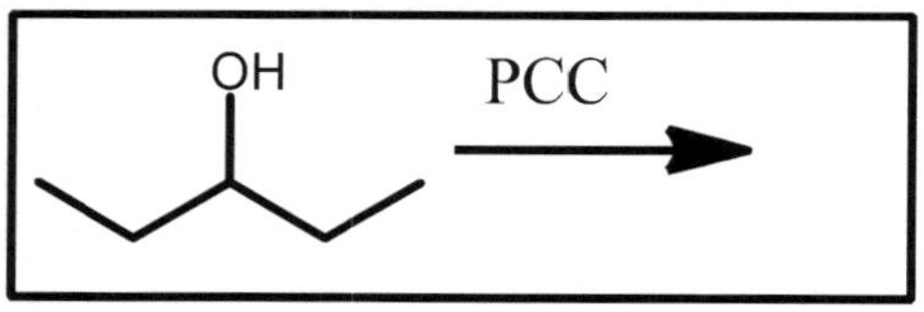

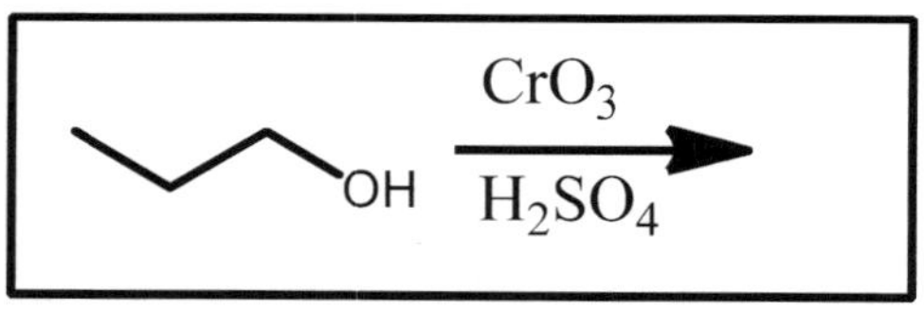

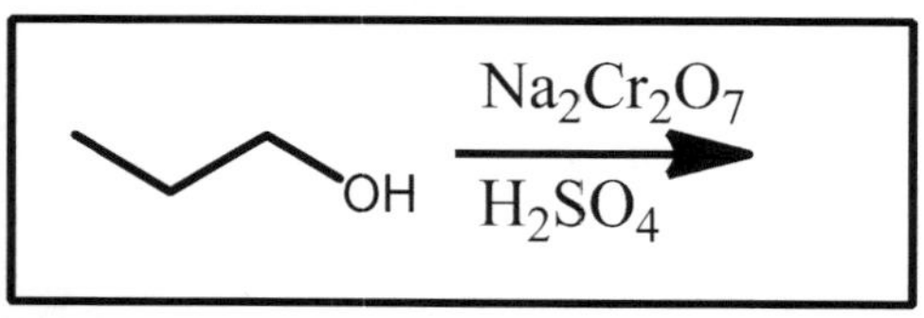

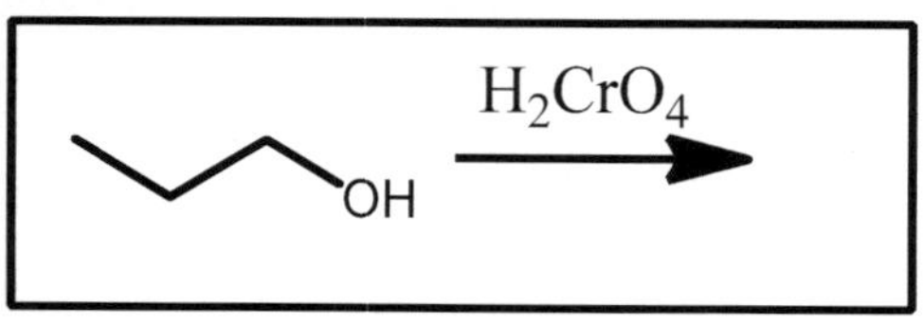

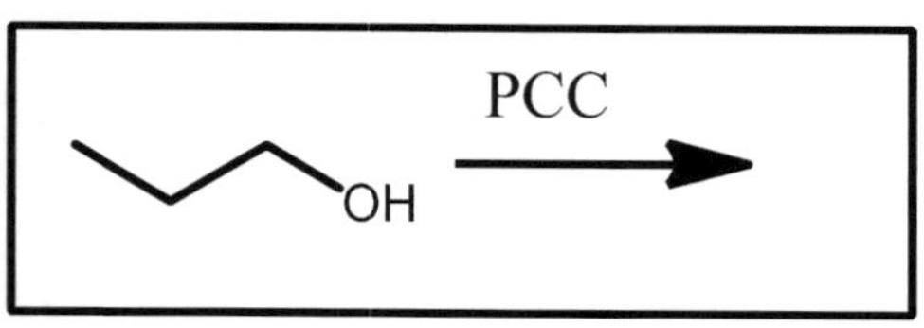

Problem Set 2: Epoxides and other Ethers

Date(s) covered in class:

Date by which material should be studied:

Chapter(s) in text covering material:

Date this problem set is due:

Points this problem set is worth:

1. Provide the major product(s) of the following reaction, including stereochemistry where relevant. Show a reasonable arrow-pushing mechanism by which the major product could form. Label intermediates and transition states where they occur in your mechanism. For concerted reactions, always draw the transition state.

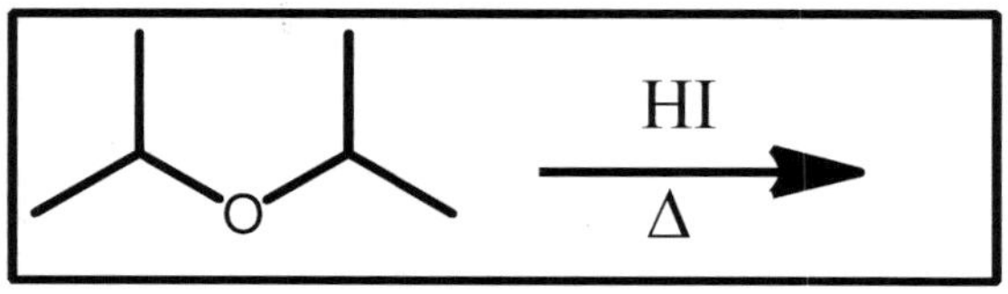

2. Provide the major product(s) of the following reaction, including stereochemistry where relevant. Show a reasonable arrow-pushing mechanism by which the major product could form. Label intermediates and transition states where they occur in your mechanism. For concerted reactions, always draw the transition state.

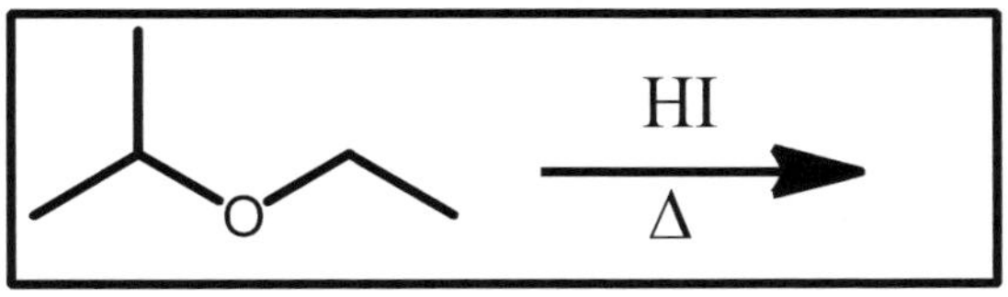

3. Provide the major product(s) of the following reaction, including stereochemistry where relevant. Show a reasonable arrow-pushing mechanism by which the major product could form. Label intermediates and transition states where they occur in your mechanism. For concerted reactions, always draw the transition state.

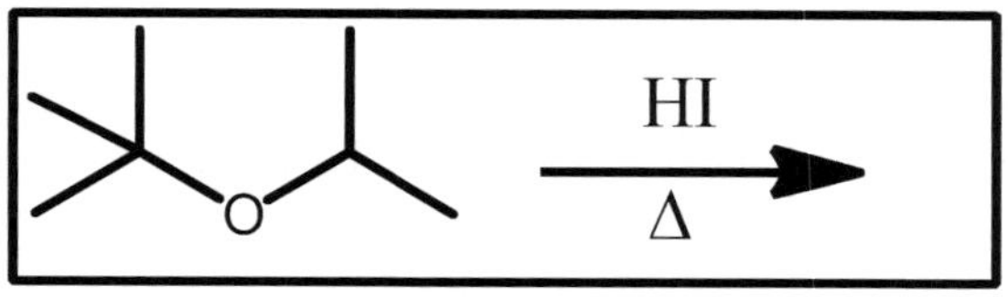

4. Provide the major product(s) of the following reaction, including stereochemistry where relevant. Show a reasonable arrow-pushing mechanism by which the major product could form. Label intermediates and transition states where they occur in your mechanism. For concerted reactions, always draw the transition state.

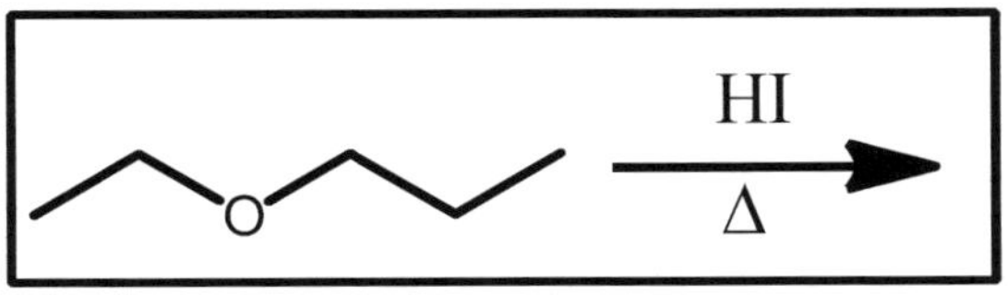

5. Provide the major product(s) of the following reaction, including stereochemistry where relevant. Show a reasonable arrow-pushing mechanism by which the major product could form. Label intermediates and transition states where they occur in your mechanism. For concerted reactions, always draw the transition state.

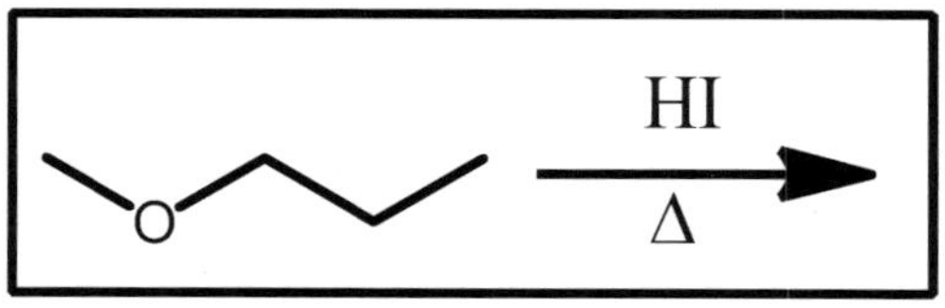

6a Provide the major product(s) of the following reaction, including stereochemistry where relevant. Show a reasonable arrow-pushing mechanism by which the major product could form. Label intermediates and transition states where they occur in your mechanism. For concerted reactions, always draw the transition state.

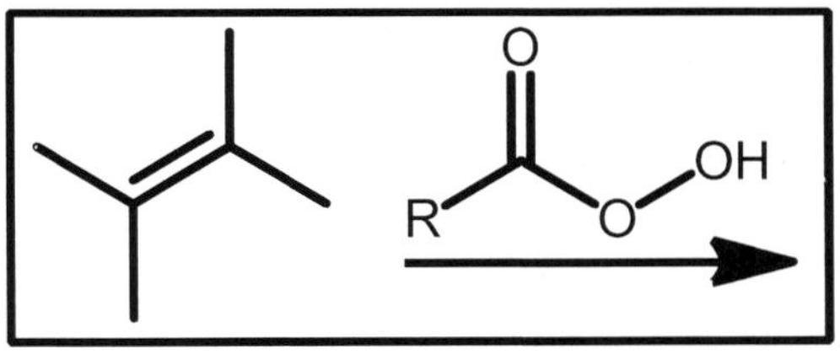

6b. If R = 3-chlorophenyl, what are the name and acronym given to the peroxy acid used in the reaction of part 6a?

7. Provide the major product(s) of the following reactions, including stereochemistry where relevant. Show a reasonable arrow-pushing mechanism by which the major product could form. Label intermediates and transition states where they occur in your mechanism. For concerted reactions, always draw the transition state.

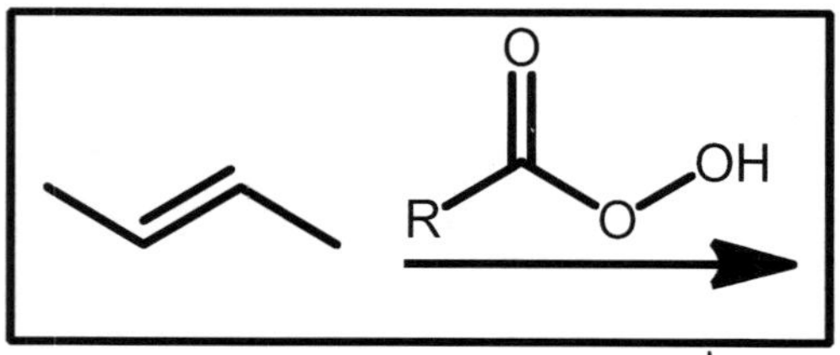

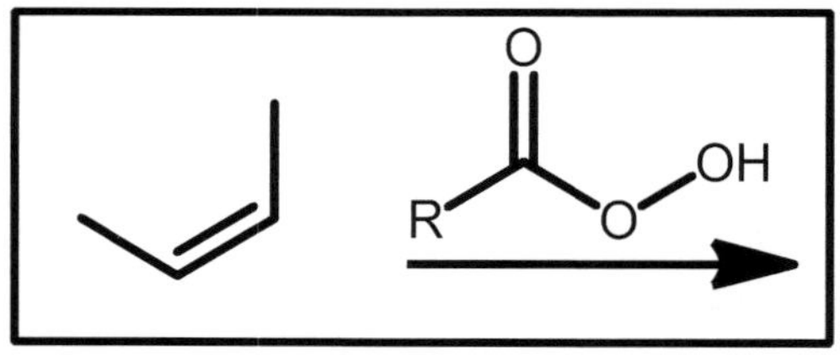

8. Sometimes, a different product is obtained from an alkene depending on which *face* of the alkene is attacked by a reagent. Provide the major product(s) of the following reactions, including the *R*- and *S*-configuration labels where relevant. Clearly indicate which face of the alkene must be attacked by the peroxy acid to yield each product.

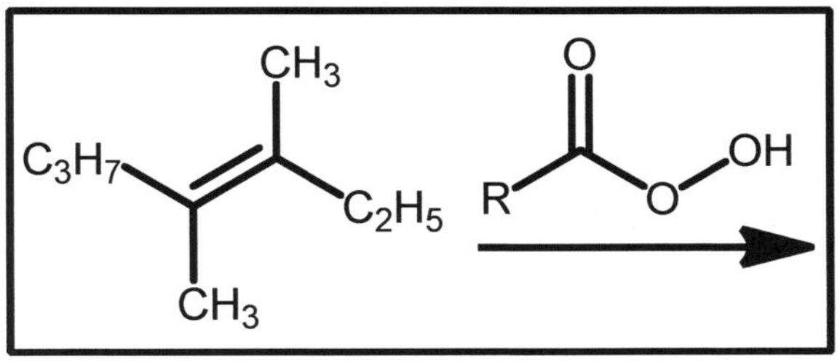

9. Provide the major product(s) of the following reactions, including stereochemistry where relevant. Show a reasonable arrow-pushing mechanism by which the major product could form. Label intermediates and transition states where they occur in your mechanism. For concerted reactions, always draw the transition state.

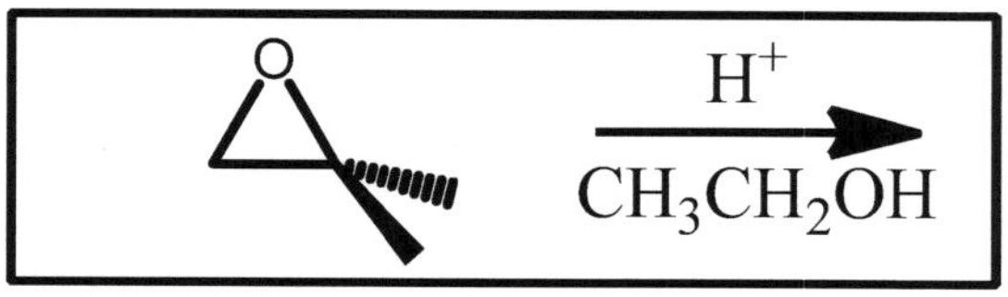

10. Provide the major product(s) of the following reactions, including stereochemistry where relevant. Show a reasonable arrow-pushing mechanism by which the major product could form. Label intermediates and transition states where they occur in your mechanism. For concerted reactions, always draw the transition state.

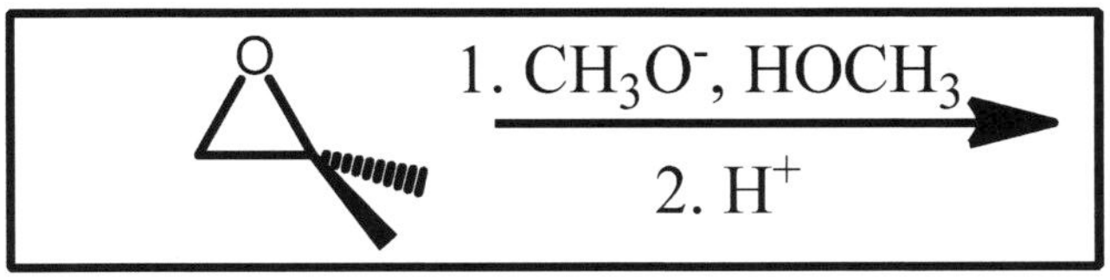

11. Provide the major product(s) of the following reactions, including stereochemistry where relevant.

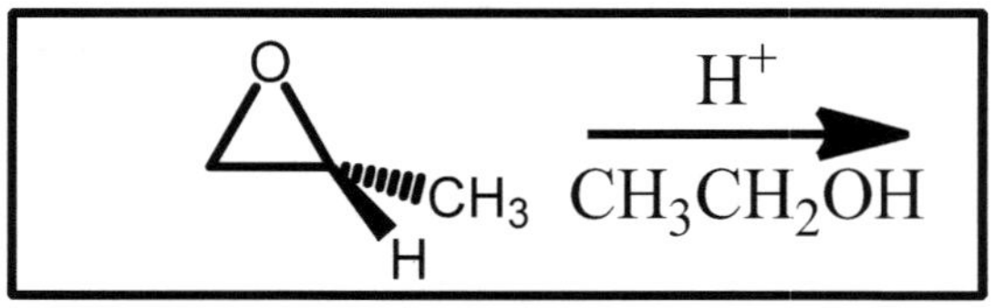

12. Provide the major product(s) of the following reactions, including stereochemistry where relevant.

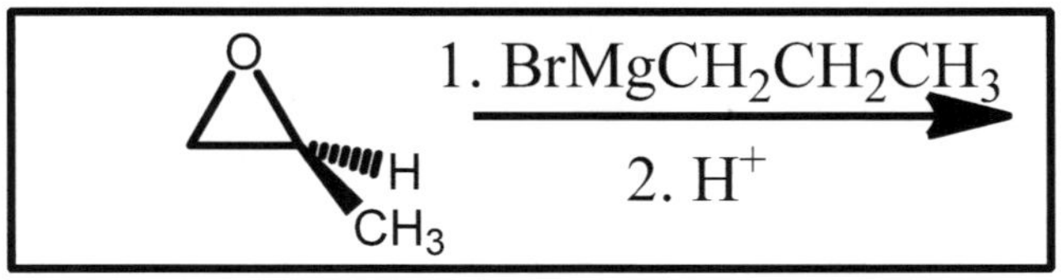

13. Propose an arrow-pushing mechanism for this reaction to form a cyclic ether:

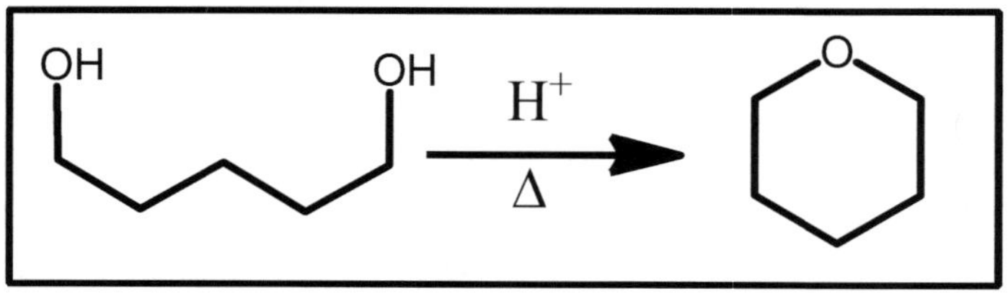

14. Provide the correct chemical names for the following molecules. Include stereochemical labels where necessary.

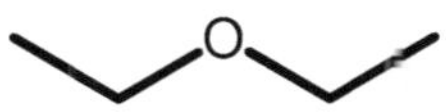

____

____

____

____

____

15. Propose an arrow-pushing mechanism for this reaction of an epoxide. LDA = lithium diisopropylamide (LiN^iPr_2):

16. Provide the major product(s) of the following reactions, including stereochemistry where relevant. Show a reasonable arrow-pushing mechanism by which the major product could form. Label intermediates and transition states where they occur in your mechanism. For concerted reactions, always draw the transition state.

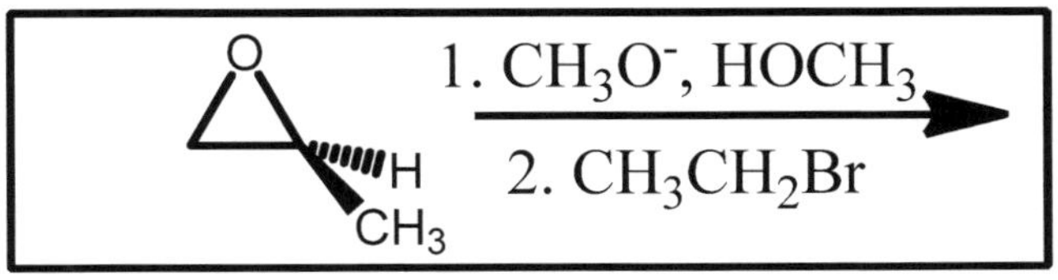

Problem Set 3: Organometallic Reagents and Multistep Synthesis

Date(s) covered in class:

Date by which material should be studied:

Chapter(s) in text covering material:

Date this problem set is due:

Points this problem set is worth:

1. This question deals with organomagnesium reagents

1a. What is the name given to an organomagnesium reagent of the form R-Mg-X (X = a halogen)?

1b. What reagents are needed to prepare the following reagent?

1c. What is the name of the general reaction wherein a metal inserts into a bond with concomitant increase in the metal's oxidation state by two units of charge?

1d. Comment on the basicity and nucleophilicity of RMgX compared to other bases and nucleophiles that have been covered in the course.

2. This question deals with organocuprate reagents

2a. What is the name given to an organocuprate reagent of the form $Li(CuR_2)$?

2b. What reagents are needed to prepare the following reagent?

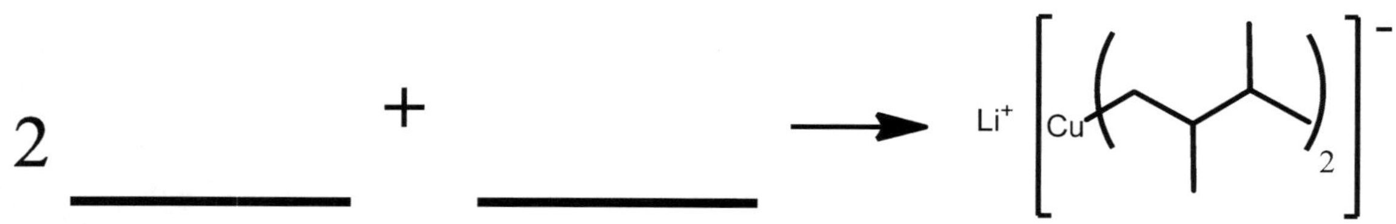

2c. Comment on the relative reactivity of the $Li(CuR_2)$ reagents as compared to RMgX or RLi reagents.

3. Provide a catalytic cycle for the Heck reaction between styrene and iodobenzene. Label any step in the catalytic cycle that is an oxidative addition, reductive elimination or transmetallation step.

4. Provide a catalytic cycle for the Suzuki reaction between phenylboronic acid and iodobenzene. Label any step in the catalytic cycle that is an oxidative addition, reductive elimination or transmetallation step.

5. Provide a catalytic cycle for the Stille reaction between tributylphenyltin and iodobenzene. Label any step in the catalytic cycle that is an oxidative addition, reductive elimination or transmetallation step.

6. a) What name is given to this reaction?

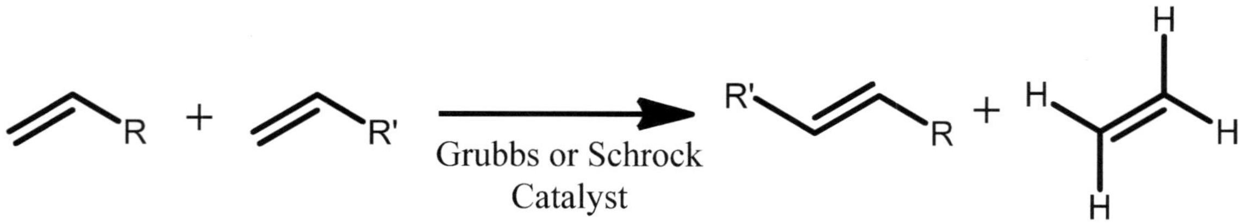

b) Provide a mechanism, representing the metal center of the Grubbs or Schrock catalyst used as M.

c) Provide an example of Alkyne Metathesis. What two metals are the most common ones in the catalysts used for this transformation?

d. Provide an example of a ROMP (ring-opening metathesis polymerization) reaction.

7. Fill in the reagents needed to carry out the following transformations.

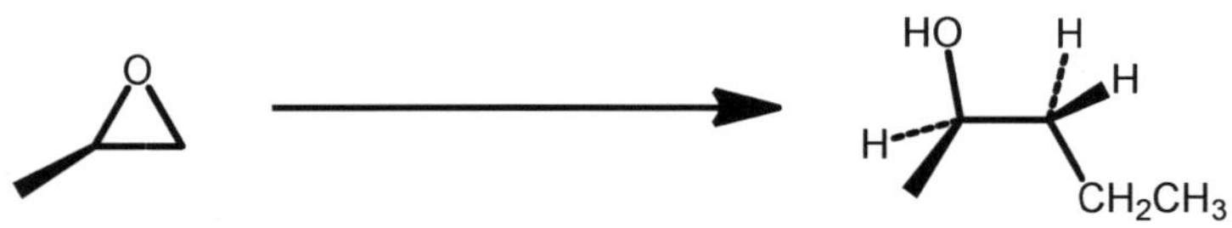

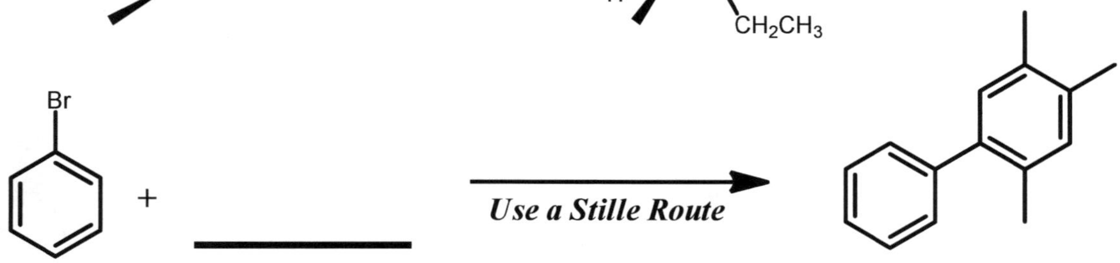

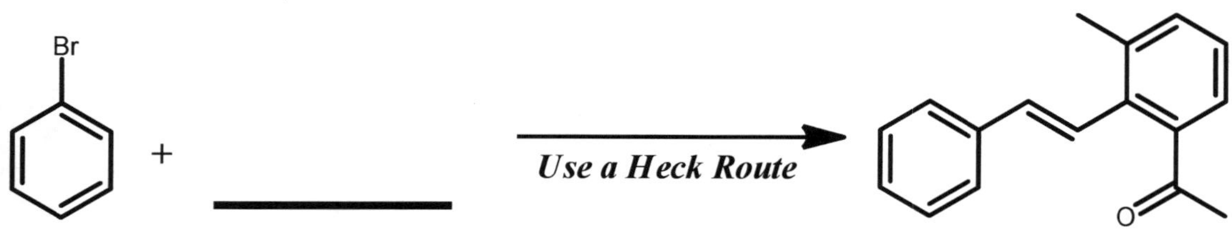

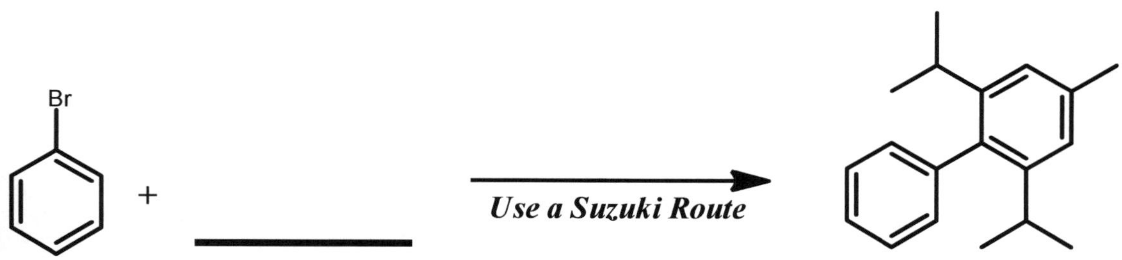

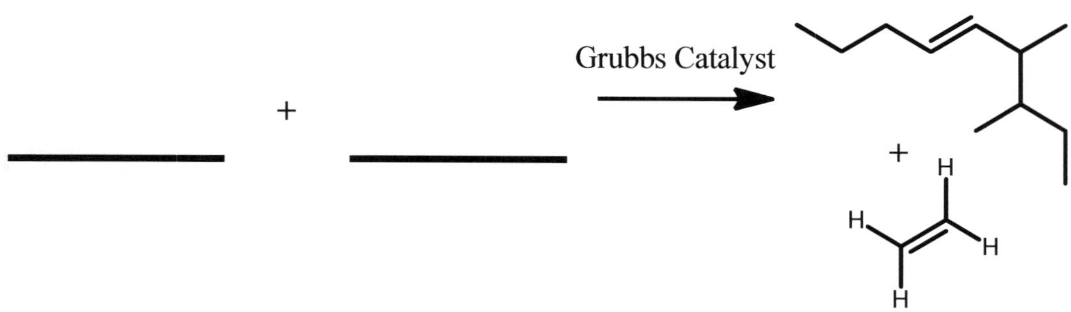

8. Provide all of the major products, both organic and inorganic, for the following reactions. Specify stereochemistry where relevant.

$$CH_3CH_2MgX \quad + \quad HOCH_2CH_2CH_3 \longrightarrow$$

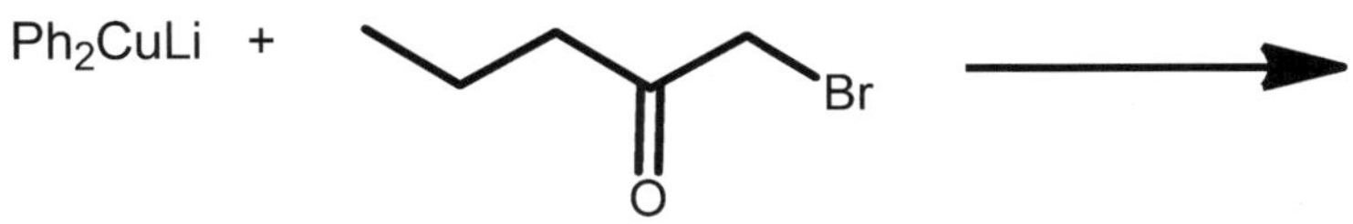

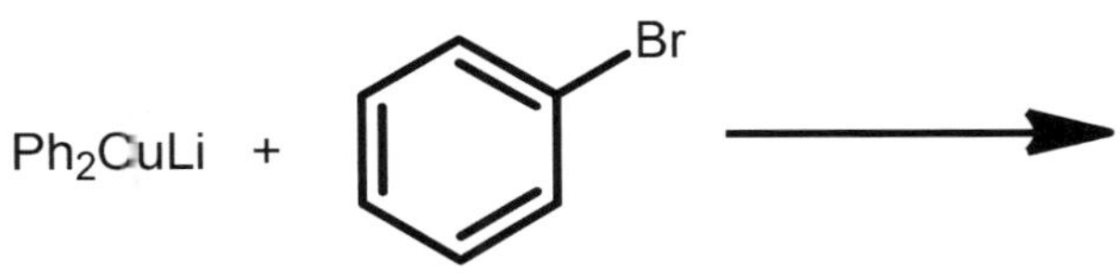

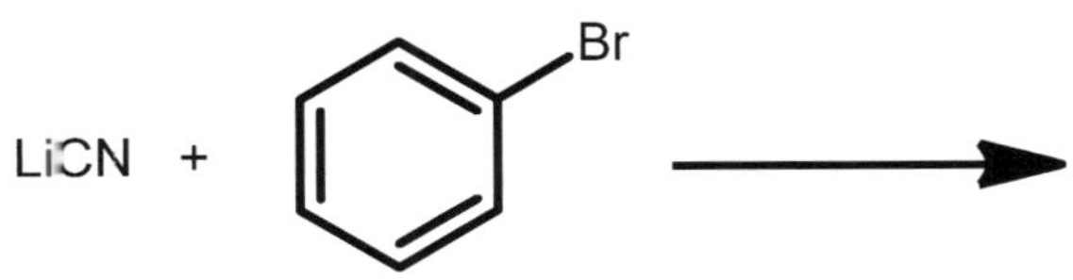

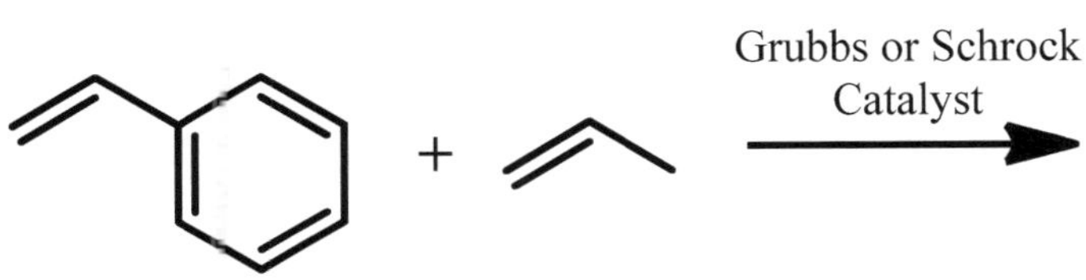

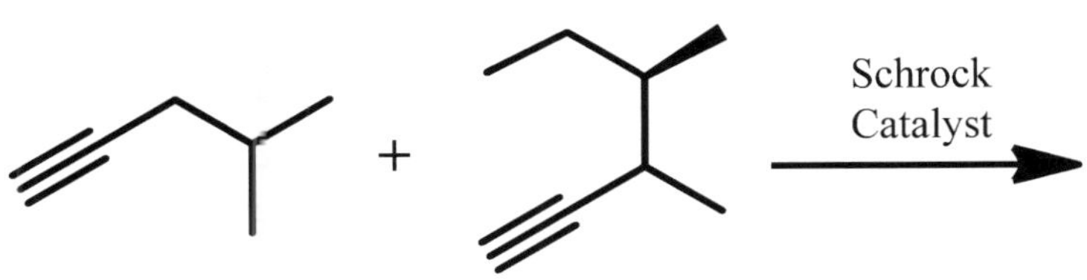

9. Propose a route to make each of the following products from the provided starting material. Use as many steps as necessary. In addition to the starting materials provided you may use any catalysts, and any necessary reagents with ≤3 carbons in its structure.

9a.

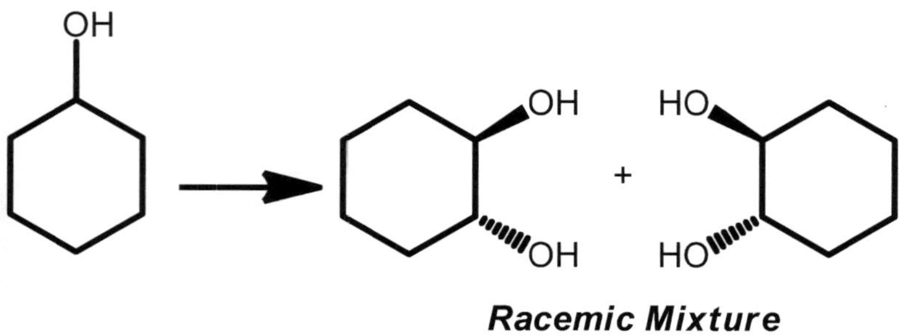

9b.

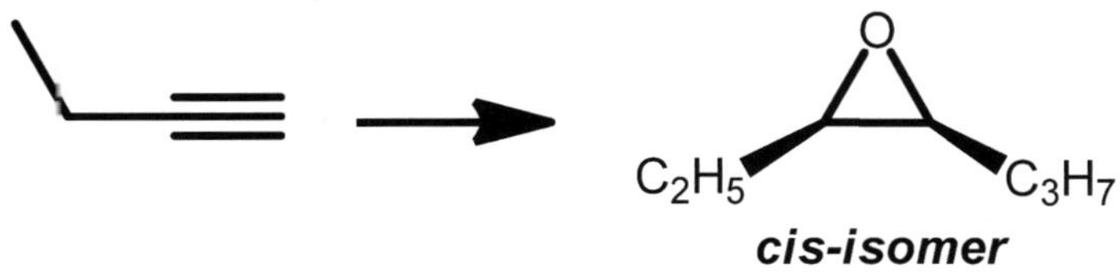

9c.

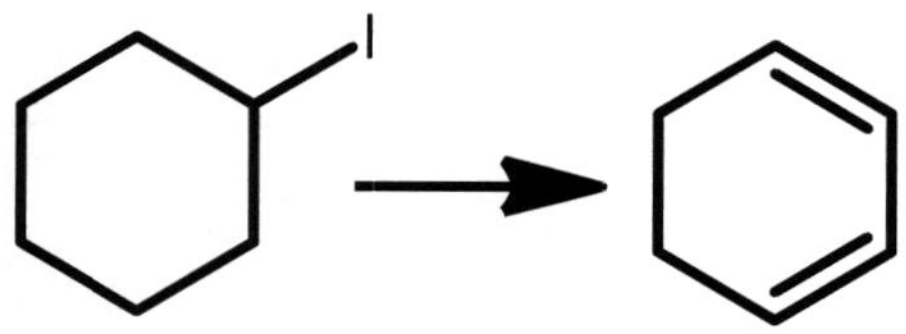

9d.

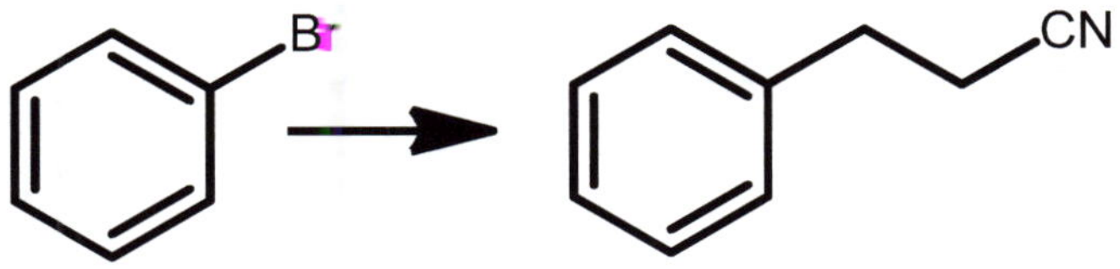

9e.

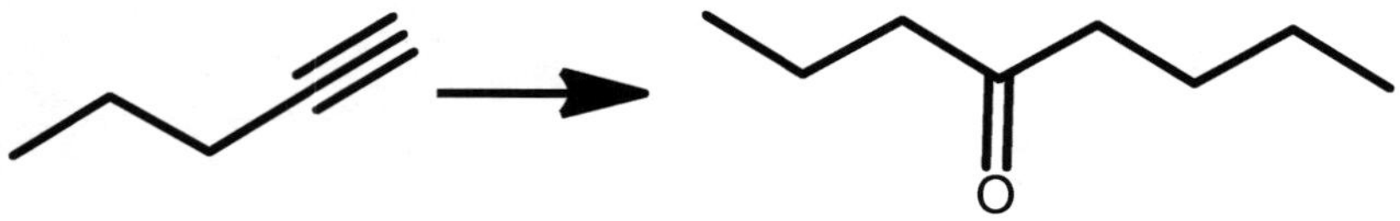

9f.

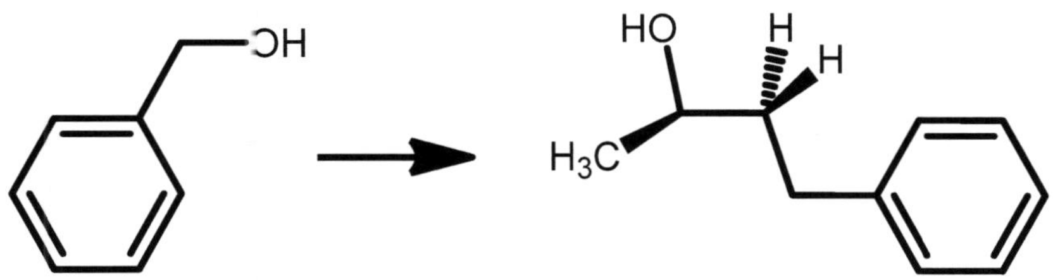

9g.

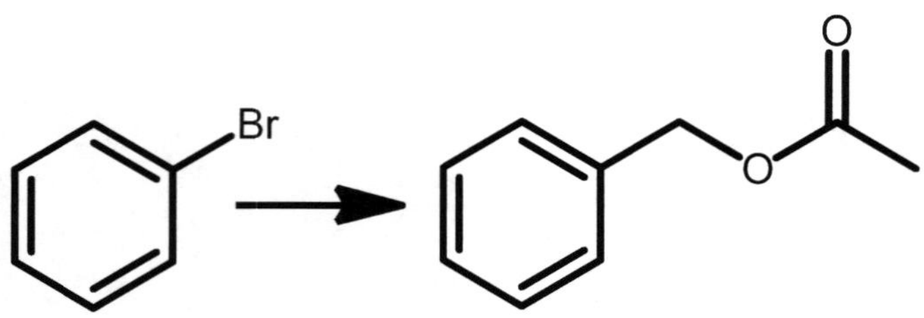

9h.

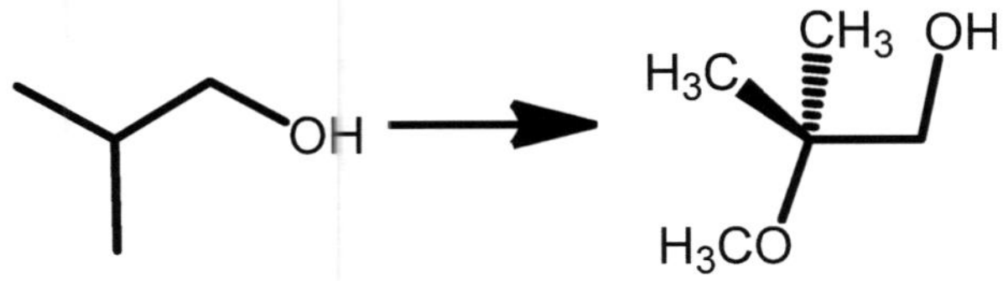

9i.

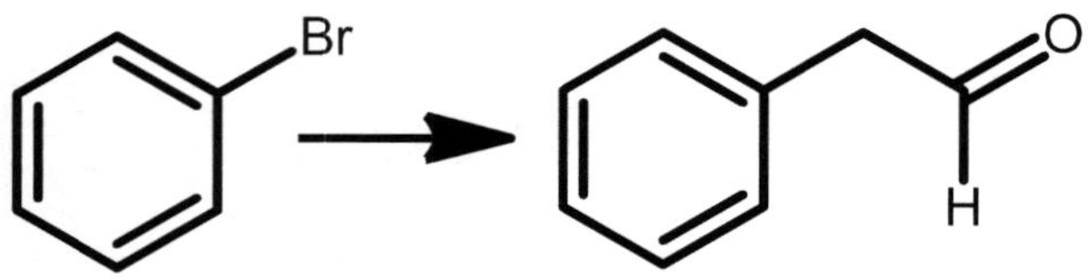

9j.

trans-isomer

9k.

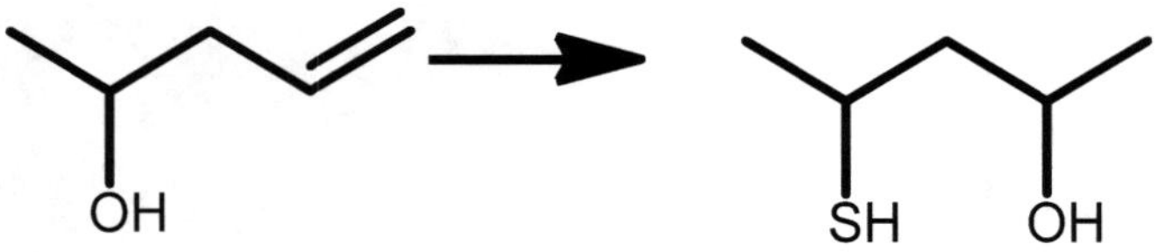

91.

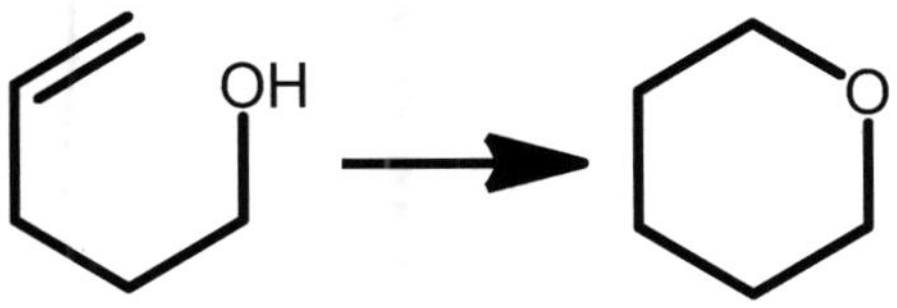

Problem Set 4: Aromaticity, Benzene and Electrophilic Aromatic Substitution on Benzene

Date(s) covered in class:

Date by which material should be studied:

Chapter(s) in text covering material:

Date this problem set is due:

Points this problem set is worth:

1. Provide the criteria for aromaticity. Give an example of an aromatic hydrocarbon that is not shown in your text.

2. How many p electrons are present in each of the following structures? Label each structure as aromatic, anti-aromatic, or neither.

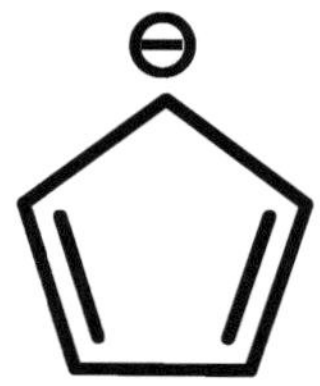

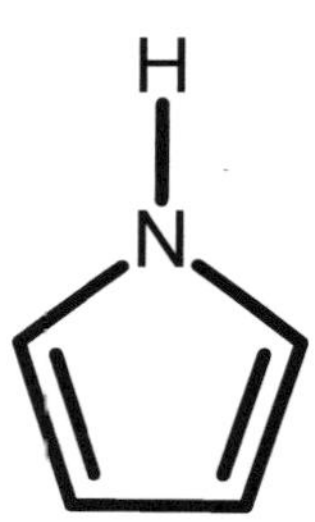
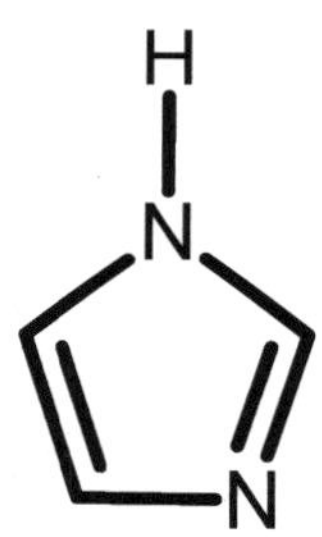
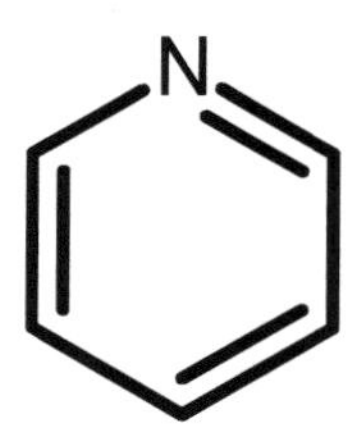

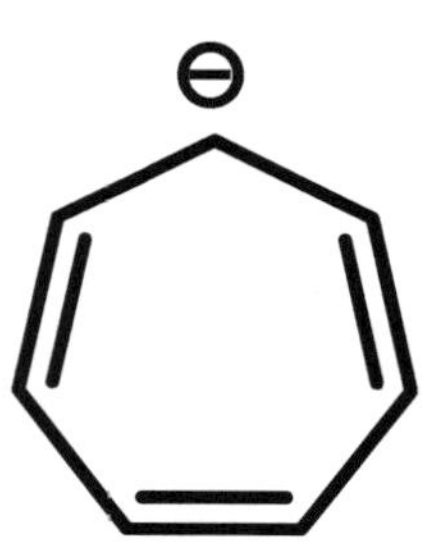
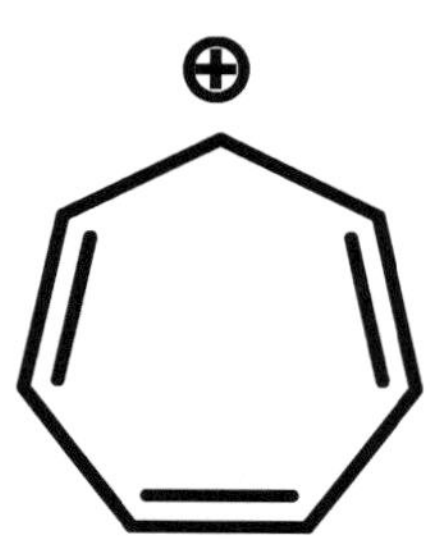

3. Label the most basic N in each compound. Identify which compound is the weakest base and which is the strongest. Justify your choices by sketching orbital pictures.

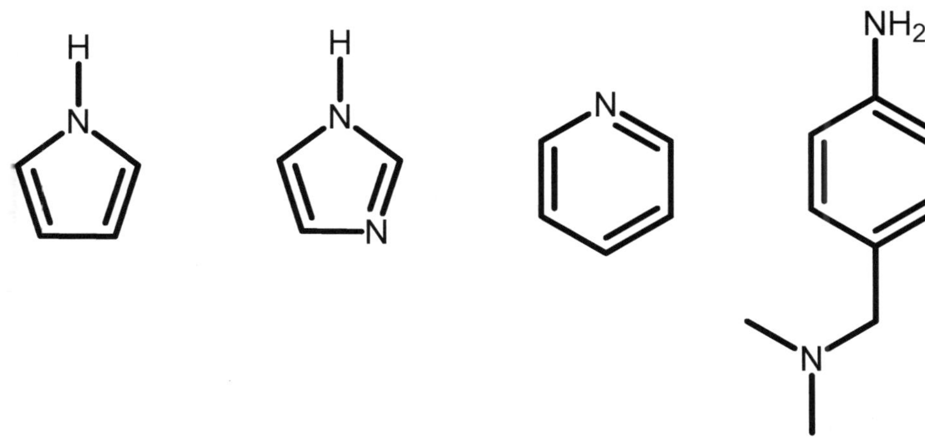

4. Provide the major product(s) of the following reaction, including stereochemistry where relevant. Show a reasonable arrow-pushing mechanism by which the major product could form. Label intermediates and transition states where they occur in your mechanism. For concerted reactions, always draw the transition state. Be sure to include the mechanism for the formation of the electrophile.

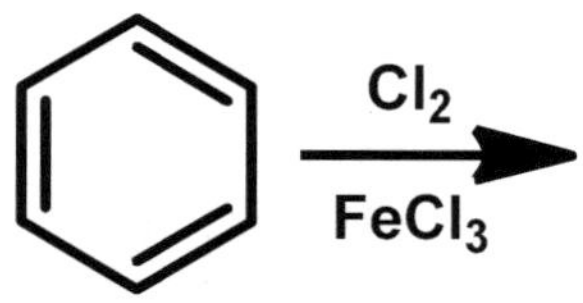

5. Provide the major product(s) of the following reaction, including stereochemistry where relevant. Show a reasonable arrow-pushing mechanism by which the major product could form. Label intermediates and transition states where they occur in your mechanism. For concerted reactions, always draw the transition state. Be sure to include the mechanism for the formation of the electrophile.

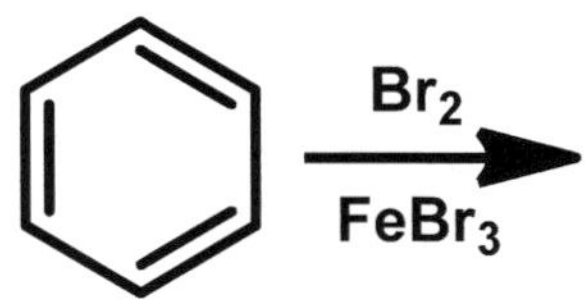

6. Provide the major product(s) of the following reaction, including stereochemistry where relevant. Show a reasonable arrow-pushing mechanism by which the major product could form. Label intermediates and transition states where they occur in your mechanism. For concerted reactions, always draw the transition state. Be sure to include the mechanism for the formation of the electrophile.

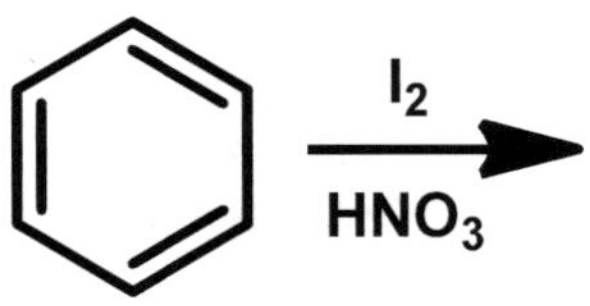

7. Provide the major product(s) of the following reaction, including stereochemistry where relevant. Show a reasonable arrow-pushing mechanism by which the major product could form. Label intermediates and transition states where they occur in your mechanism. For concerted reactions, always draw the transition state. Be sure to include the mechanism for the formation of the electrophile.

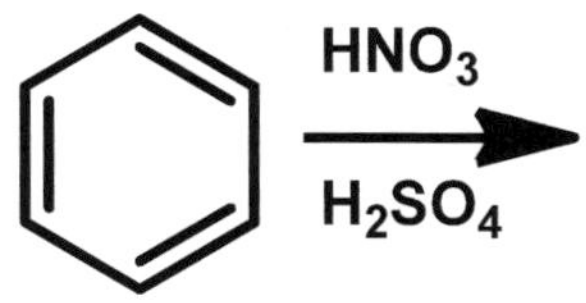

8. Provide the major product(s) of the following reaction, including stereochemistry where relevant. Show a reasonable arrow-pushing mechanism by which the major product could form. Label intermediates and transition states where they occur in your mechanism. For concerted reactions, always draw the transition state. Be sure to include the mechanism for the formation of the electrophile.

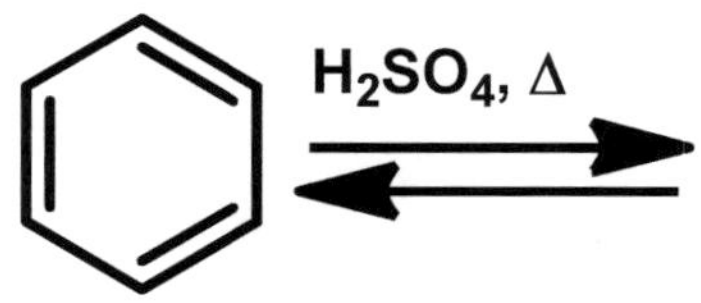

9. Provide the major product(s) of the following reaction, including stereochemistry where relevant. Show a reasonable arrow-pushing mechanism by which the major product could form. Label intermediates and transition states where they occur in your mechanism. For concerted reactions, always draw the transition state. Be sure to include the mechanism for the formation of the electrophile.

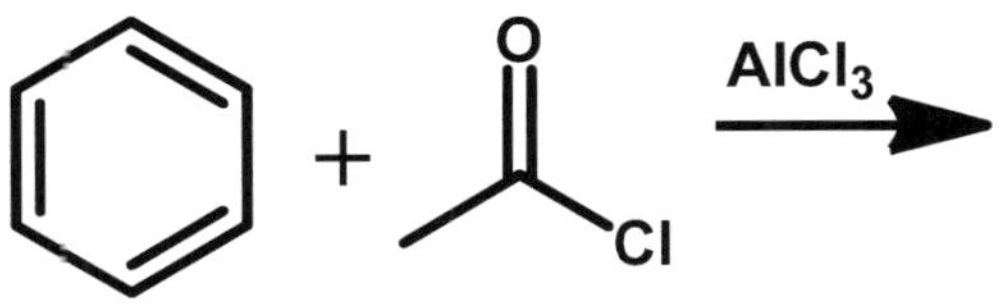

10. Provide the major product(s) of the following reaction, including stereochemistry where relevant. Show a reasonable arrow-pushing mechanism by which the major product could form. Label intermediates and transition states where they occur in your mechanism. For concerted reactions, always draw the transition state. Be sure to include the mechanism for the formation of the electrophile.

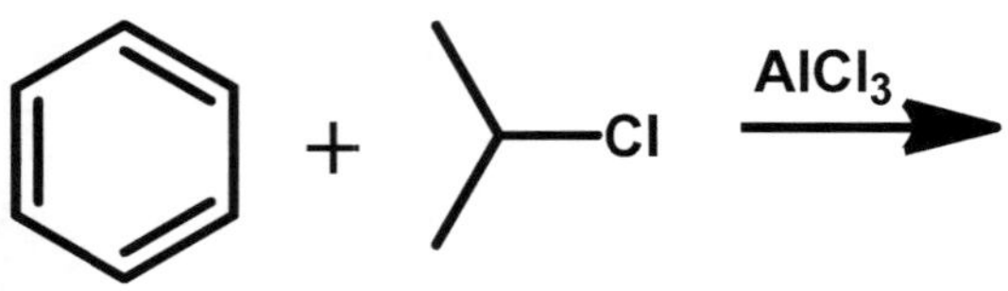

11. Provide the major product(s) of the following reaction, including stereochemistry where relevant. Show a reasonable arrow-pushing mechanism by which the major product could form. Label intermediates and transition states where they occur in your mechanism. For concerted reactions, always draw the transition state. Be sure to include the mechanism for the formation of the electrophile.

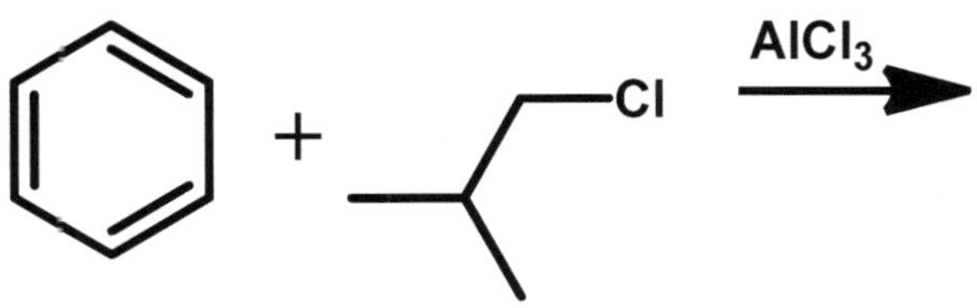

12. Provide the major product(s) of the following reaction, including stereochemistry where relevant. Show a reasonable arrow-pushing mechanism by which the major product could form. Label intermediates and transition states where they occur in your mechanism. For concerted reactions, always draw the transition state. Be sure to include the mechanism for the formation of the electrophile.

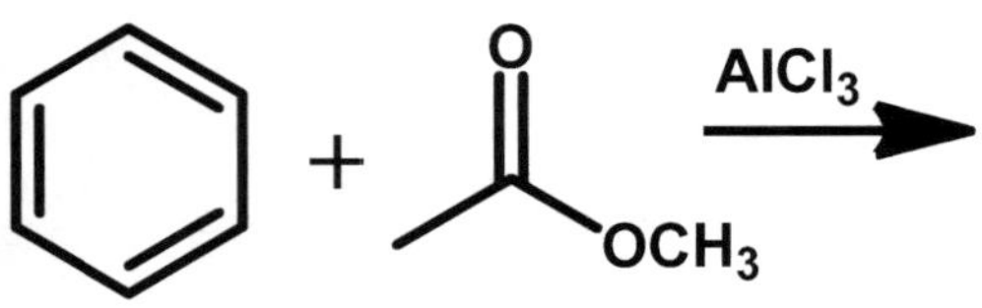

13. Provide the major product(s) of the following reaction, including stereochemistry where relevant. Show a reasonable arrow-pushing mechanism by which the major product could form. Label intermediates and transition states where they occur in your mechanism. For concerted reactions, always draw the transition state. Be sure to include the mechanism for the formation of the electrophile.

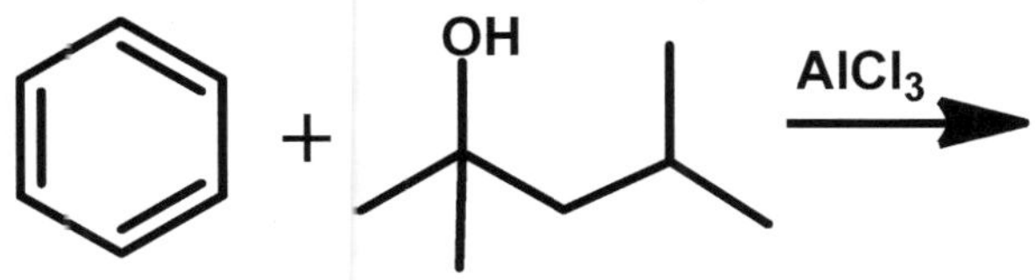

14. Provide the major product(s) of the following reaction, including stereochemistry where relevant. Show a reasonable arrow-pushing mechanism by which the major product could form. Label intermediates and transition states where they occur in your mechanism. For concerted reactions, always draw the transition state. Be sure to include the mechanism for the formation of the electrophile.

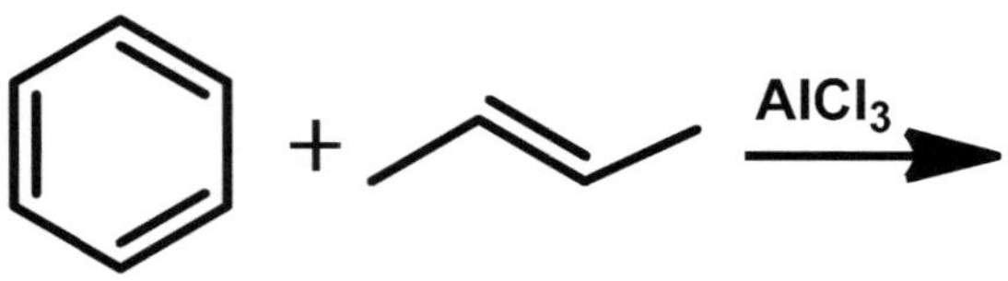

15. Provide the major product(s) of the following reaction, including stereochemistry where relevant. Show a reasonable arrow-pushing mechanism by which the major product could form. Label intermediates and transition states where they occur in your mechanism. For concerted reactions, always draw the transition state. Be sure to include the mechanism for the formation of the electrophile.

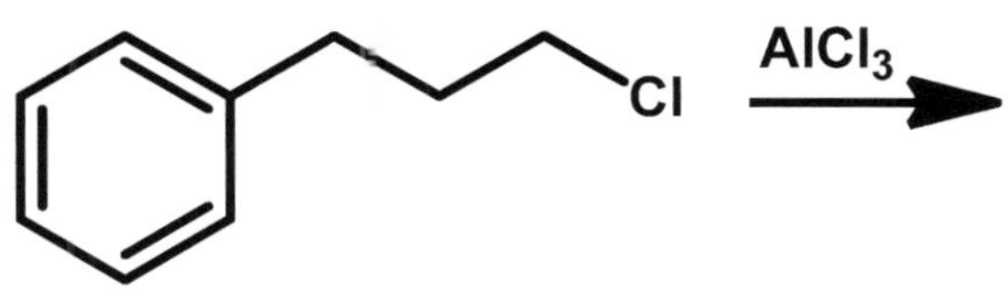

16. Provide the major product(s) of the following reaction, including stereochemistry where relevant. Show a reasonable arrow-pushing mechanism by which the major product could form. Label intermediates and transition states where they occur in your mechanism. For concerted reactions, always draw the transition state. Be sure to include the mechanism for the formation of the electrophile.

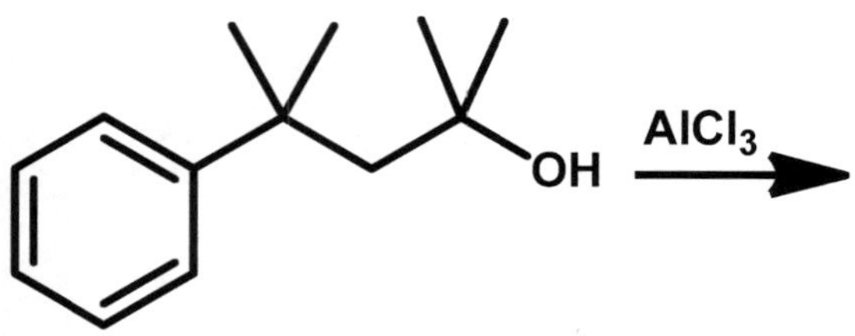

17. Provide the major product(s) of the following reaction, including stereochemistry where relevant. Show a reasonable arrow-pushing mechanism by which the major product could form. Label intermediates and transition states where they occur in your mechanism. For concerted reactions, always draw the transition state. Be sure to include the mechanism for the formation of the electrophile.

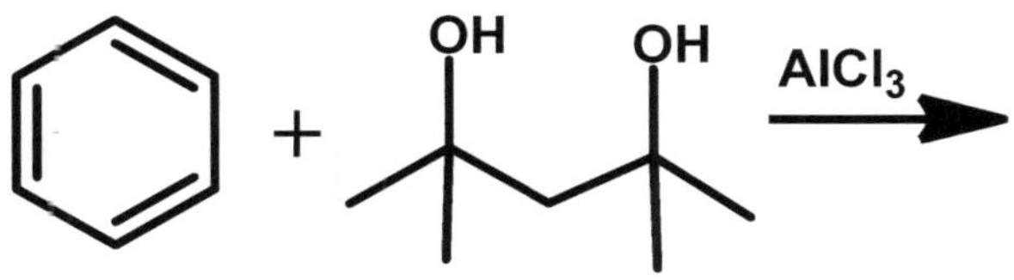

Problem Set 5: Electrophilic Aromatic Substitution on Substituted Benzene

Date(s) covered in class:

Date by which material should be studied:

Chapter(s) in text covering material:

Date this problem set is due:

Points this problem set is worth:

1. Draw the resonance structures for the initial carbocation that will form in the chlorination of *N,N*-dimethylaniline where the chloro substituent adds a) to the *ortho*- position and b) to the *meta*- position. On the basis of these resonance structures, which substitution site do you think is more favorable? Explain your reasoning.

1a)

(part b is on the next page)

1b)

2. Draw the resonance structures for the initial carbocation that will form in the chlorination of acetophenone (methyl phenyl ketone) where the chloro substituent adds a) to the *ortho-* position and b) to the *meta-* position. On the basis of these resonance structures, which substitution site do you think is more favorable? Explain your reasoning.

2a)

(part b is on the next page)

2b)

<u>Explanation:</u>

3. Rank these substituents in terms of their activating/deactivating nature in electrophilic aromatic substitution reactions, most activating first. Explain your rationale for each choice in terms of resonance and inductive effects.

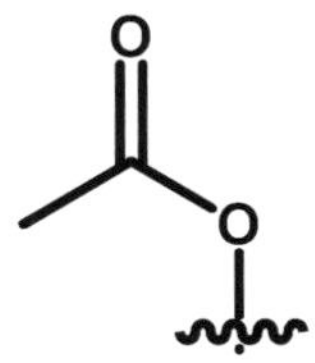

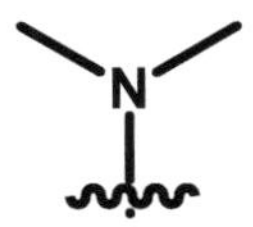

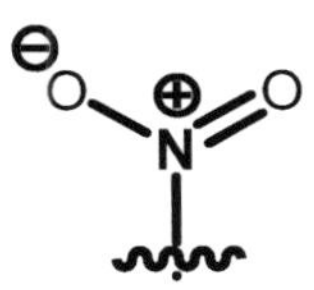

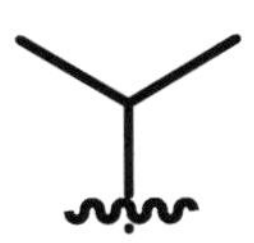

4. Provide the major product for each of the following reactions:

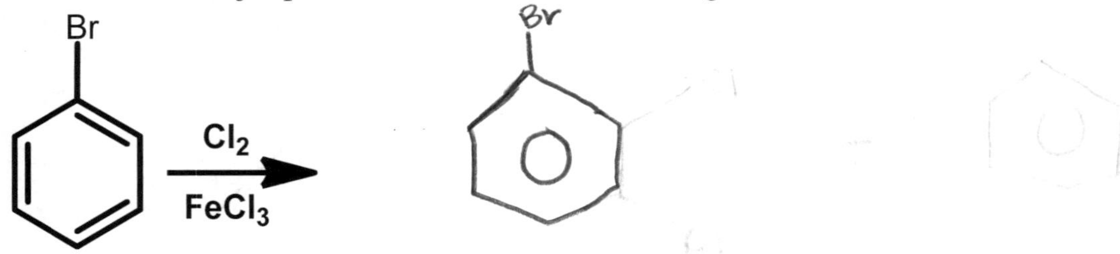

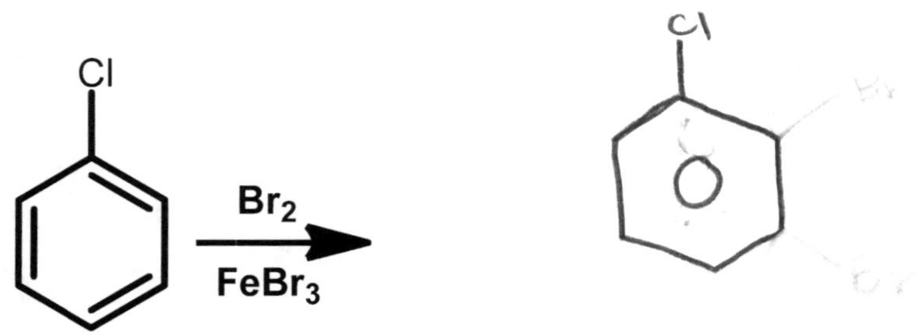

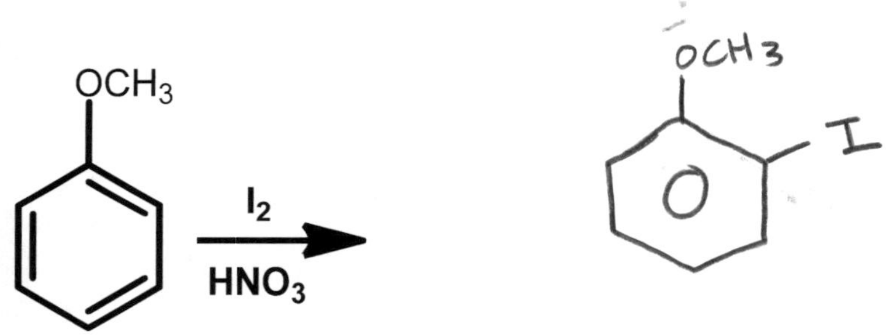

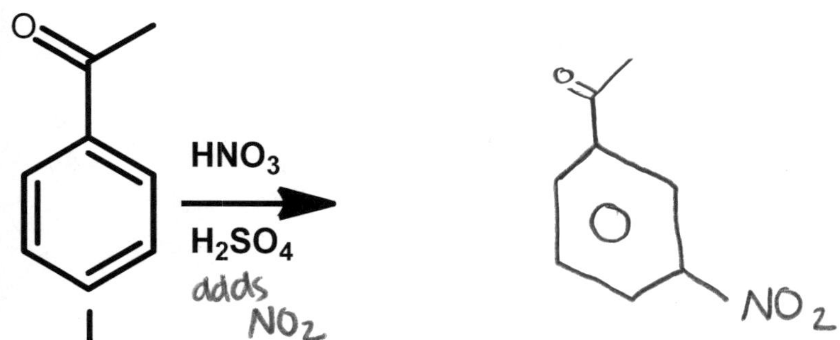

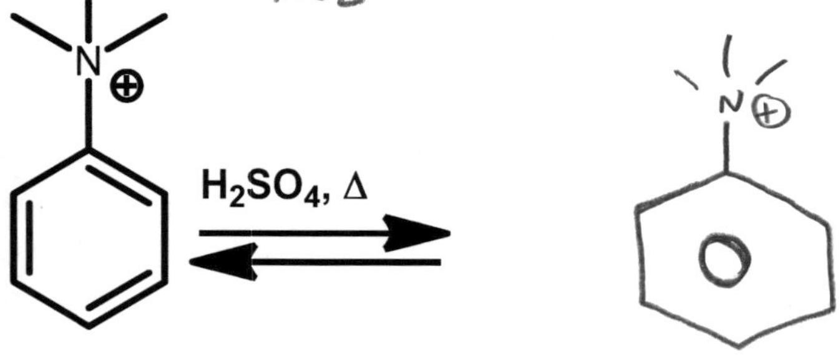

5. Provide the major product for each of the following reactions:

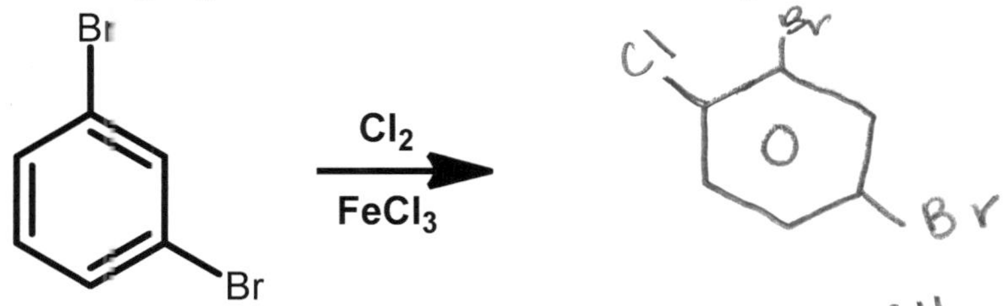

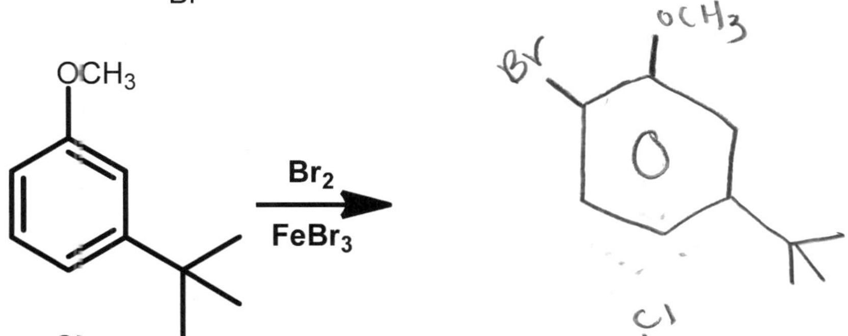

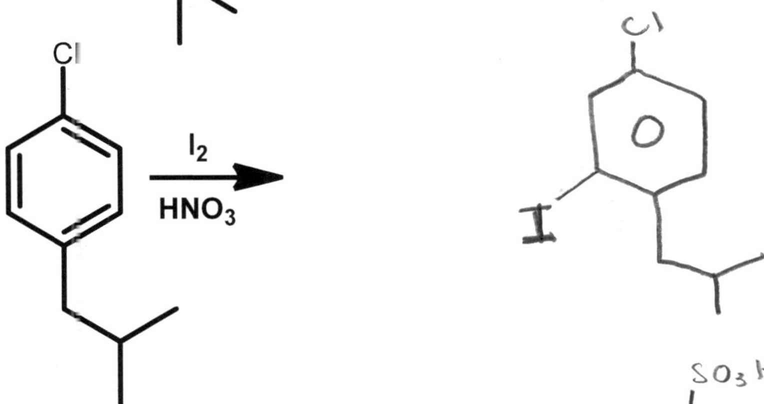

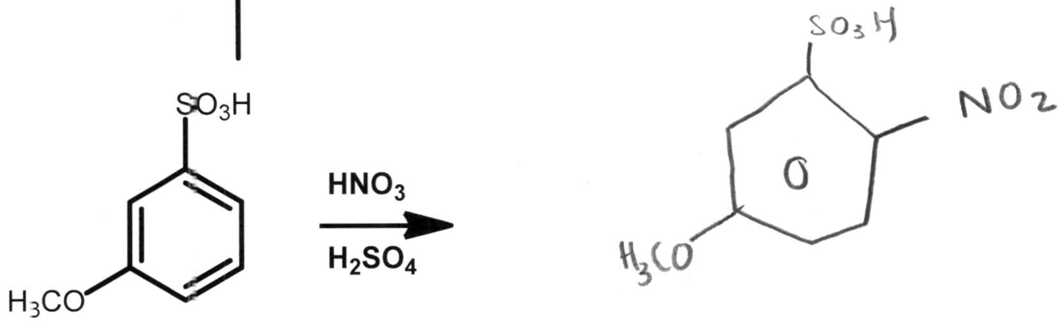

6. Provide the sequence of electrophilic aromatic substitution reactions that will give the highest yield of each of the following products (starting from benzene):

6a)

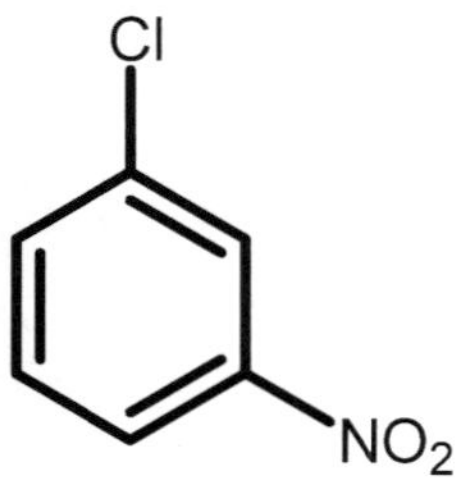

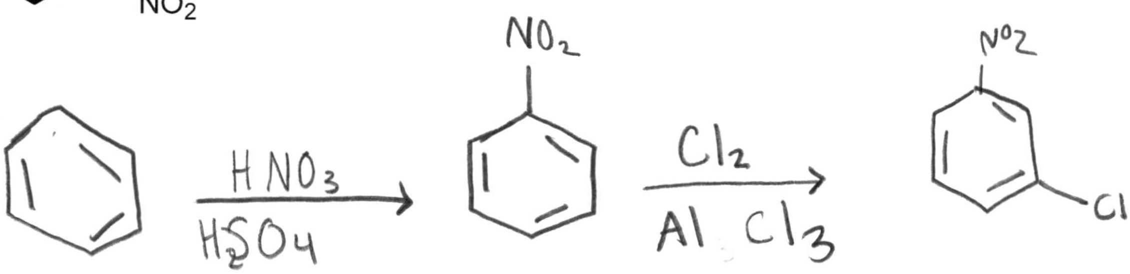

6b)

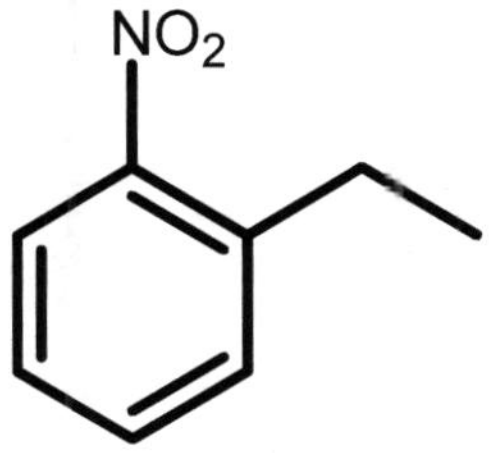

6c)

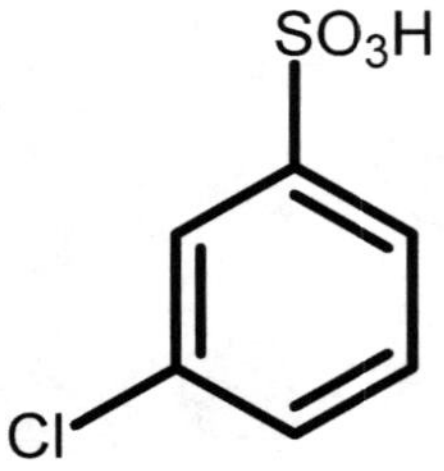

6d)

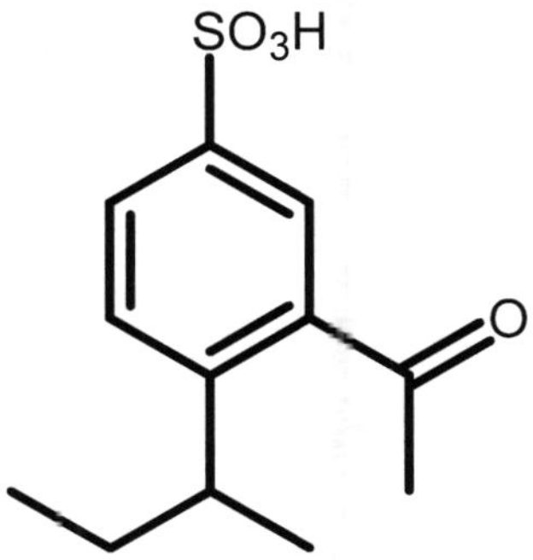

6e)

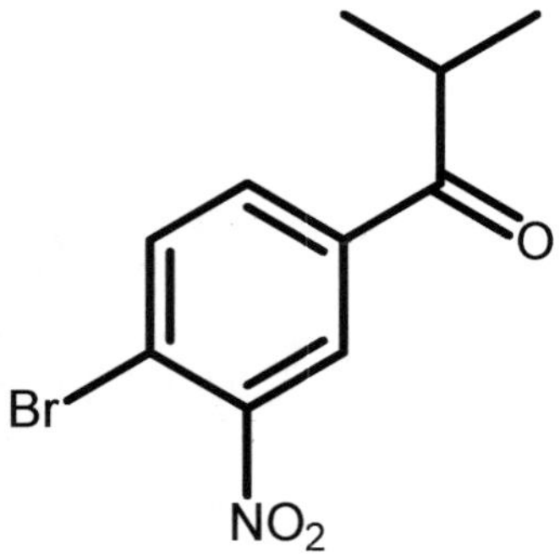

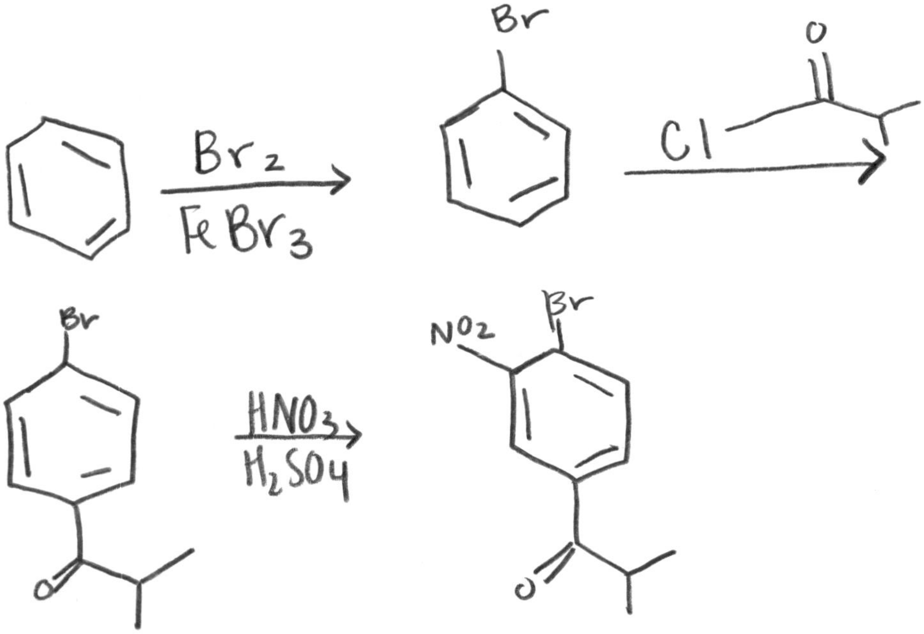

7. Identify the fastest site for electrophilic aromatic substitution on each of the following molecules (if more than one site is expected to have the same reactivity, label all equivalent sites).

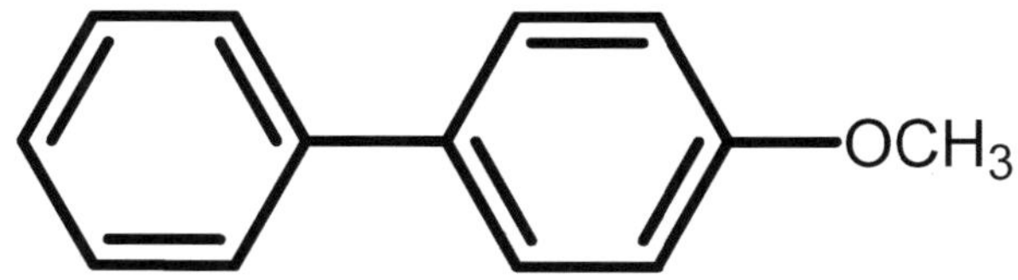

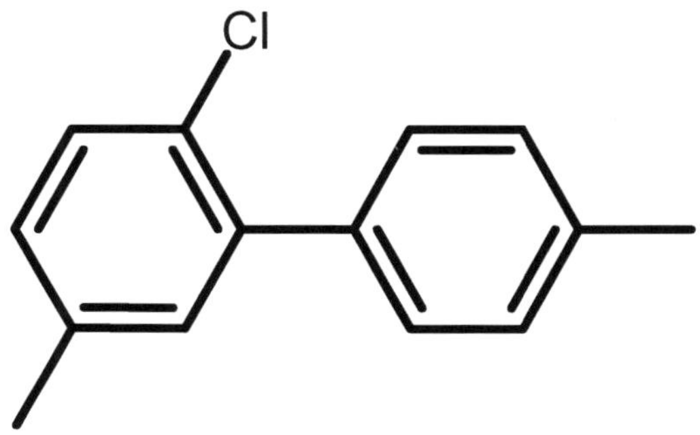

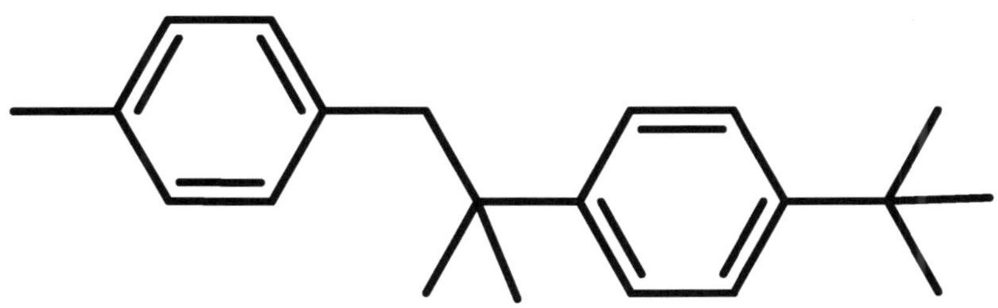

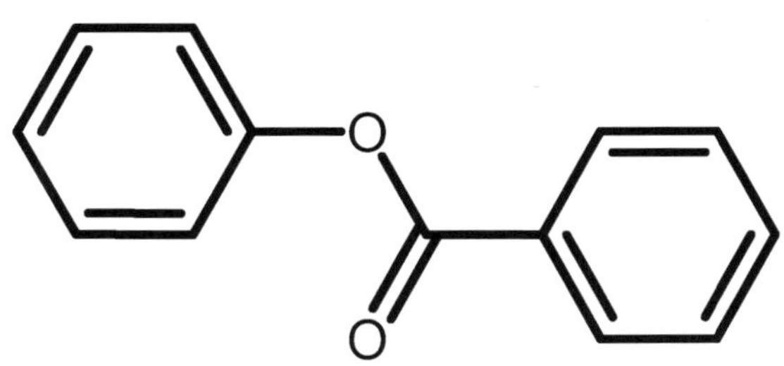

8. Using resonance structures of the intermediate carbocation that forms upon addition of the electrophile (E^+) to form each of the possible substitution products shown, predict which site is the preferred site for electrophilic aromatic substitution of naphthalene.

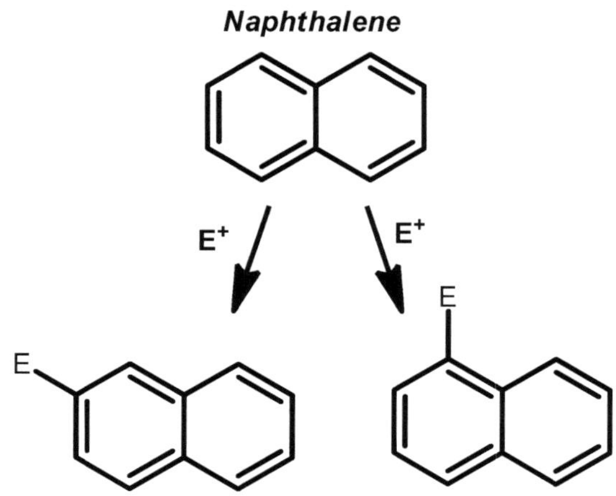

9. Predict the preferred site for electrophilic aromatic substitution on pyrrole. Will pyrrole react more or less quickly than benzene under identical conditions? Explain your reasoning.

10. Predict the preferred site for electrophilic aromatic substitution on pyridine. Will pyridine react more or less quickly than benzene under identical conditions? Explain your reasoning.

Problem Set 6: More Reactions of Arenes: Beyond Electrophilic Aromatic Substitution

Date(s) covered in class:

Date by which material should be studied:

Chapter(s) in text covering material:

Date this problem set is due:

Points this problem set is worth:

1. Provide the major product(s) of the following reaction. Then, provide an arrow-pushing mechanism for the formation of the product(s) proposed. Label the initiation, propagation, and termination steps.

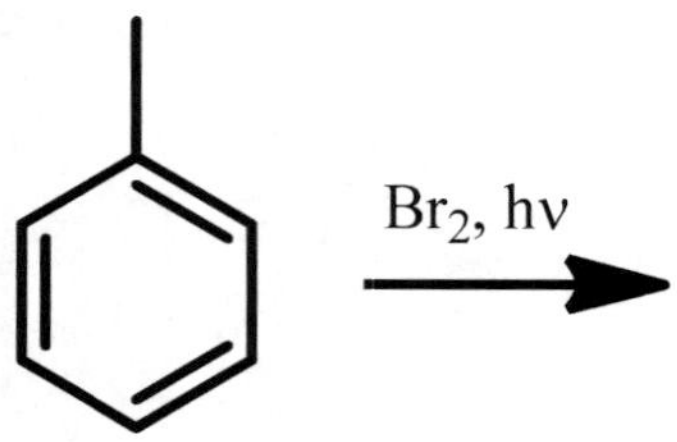

2. Provide the major product(s) of the following reaction. Draw the radicals that form from the heating of the radical initiators BPO and AIBN. What advantages might this route have over the route used in the previous problem?

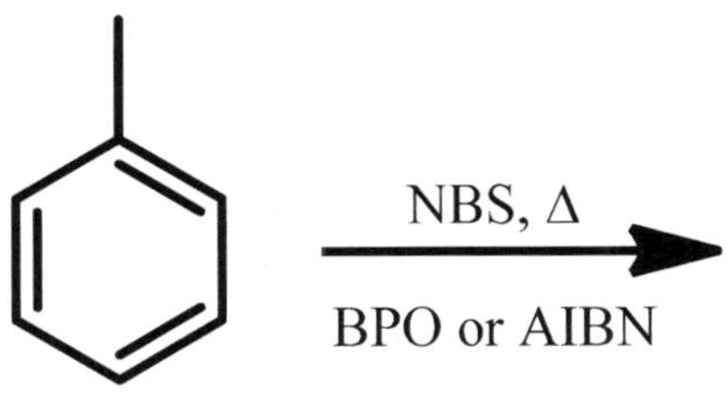

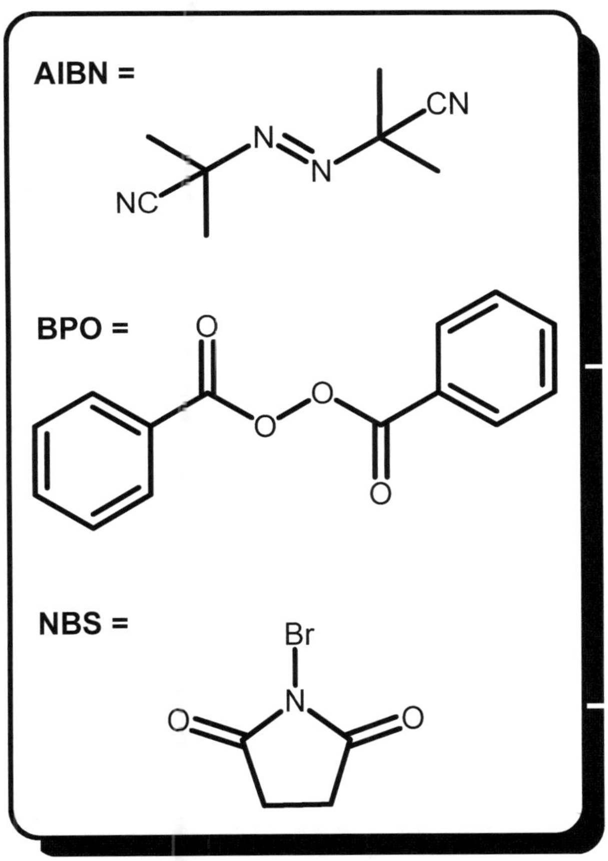

3. The benzylic position is susceptible to both S_N1 and S_N2 reactions. Provide an example of a substrate with a leaving group at the benzylic position that will proceed almost entirely by a) the S_N1; and b) almost entirely by the S_N2 mechanism.

4. The benzylic position is also susceptible to both E1 and E2 reactions. Provide an example of a substrate with a leaving group at the benzylic position that will proceed almost entirely by a) the E1; and b) almost entirely by the E2 mechanism.

5. Substituted aromatics can often be changed by **reduction** of the substituents to transform them into different substituents.

a) Provide a definition of reduction that is useful for organic molecules:

b) Above the arrows, fill in the reagent(s) needed to carry out the following reductions of substituents on a benzene ring:

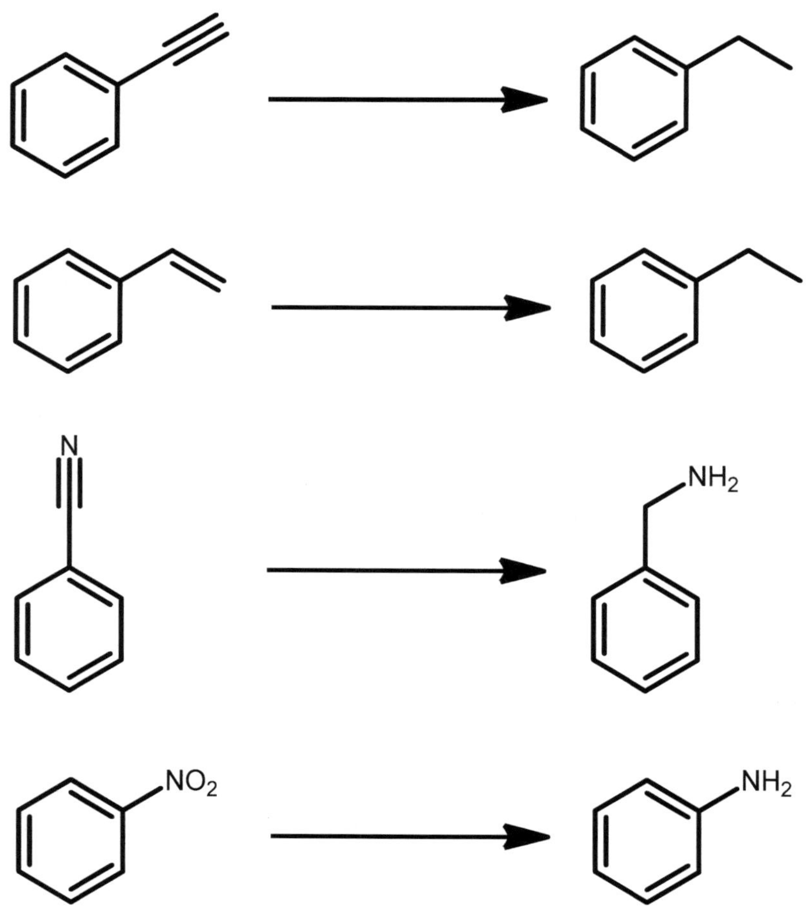

6. Although aromaticity endows the benzene ring with a great deal of stability, even the benzene ring itself can be reduced under more forcing conditions.

a) Provide the conditions needed to carry out this transformation:

b) Benzene can be reduced to a cyclohexadiene via a process known as the **Birch reduction**. This reaction involves the use of an alkali metal and liquid ammonia, conditions similar to those used in the reduction of an alkyne to a *trans*-alkene. Provide an arrow-pushing mechanism for the birch reduction given below:

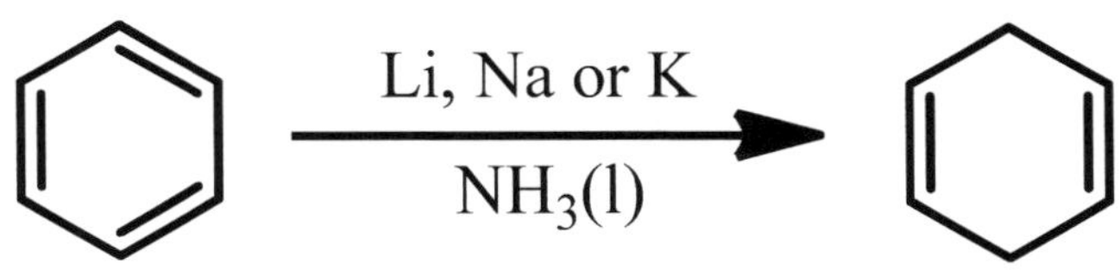

(cont'd)

7. Substituted aromatics can be changed by **oxidation** of the substituents to transform them into different substituents.

a) Provide a definition of oxidation that is useful for organic molecules:

b) Above the arrows, fill in the reagent(s) needed to carry out the following oxidations of substituents on a benzene ring. For the last one, predict the product.

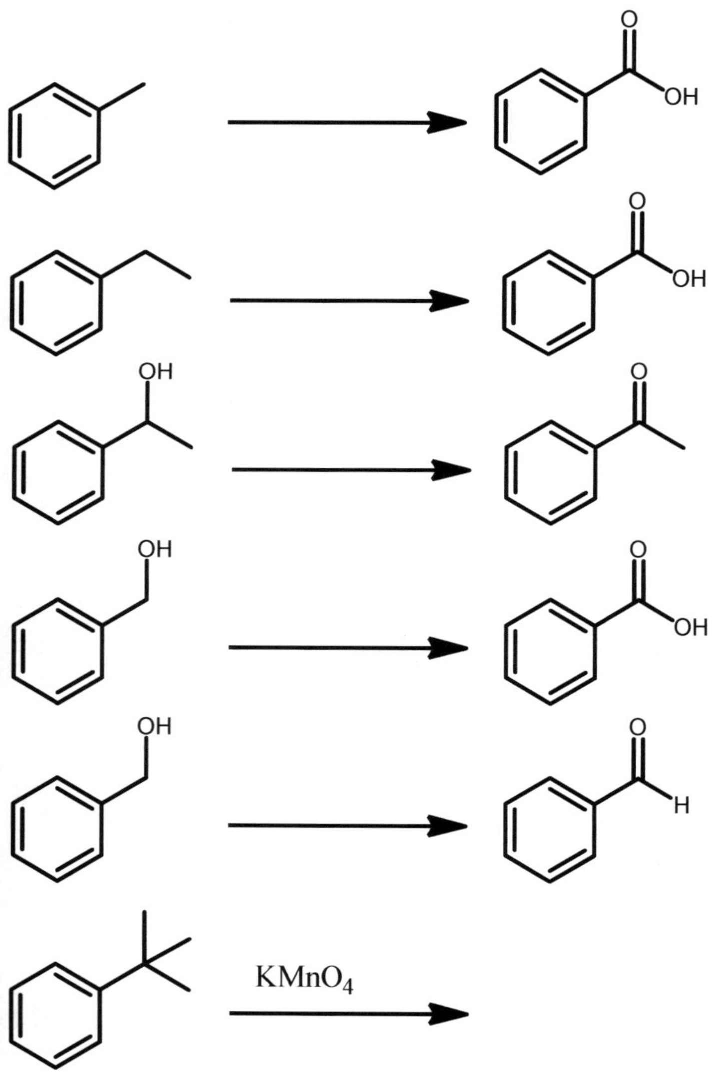

8. **Diazotization** is a very useful reaction for converting the -NH2 functionality of aniline into other groups. Provide the structure of the major product produced from the reaction of the diazonium salt with each of the provided reagents:

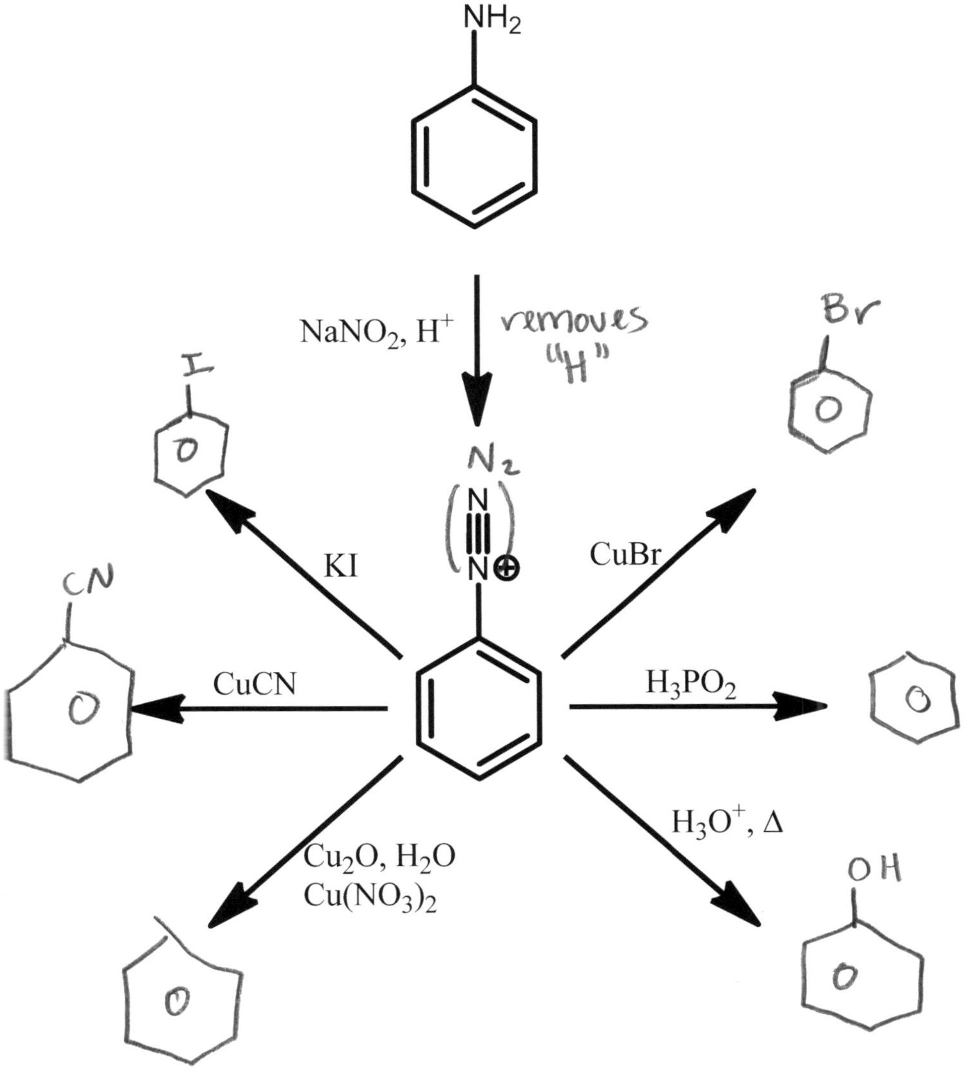

9. Because the diazonium salt is positively charged, it can actually act as an electrophile to attack activated aromatic rings in an electrophilic aromatic substitution like reaction. Provide an arrow-pushing mechanism for the following reaction:

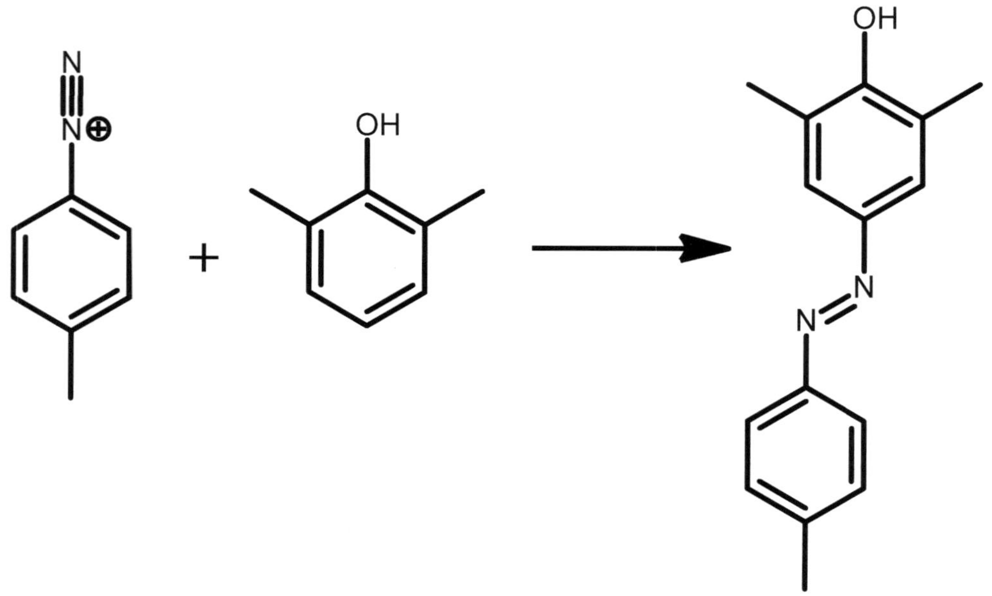

10. What are the requirements for nucleophilic aromatic substitution to take place by the **S_NAr**-type mechanism with respect to:

a) Substituents on the ring that is to be attacked by the nucleophile

b) The stability of the nucleophile compared to that of the leaving group on the ring

c) The position of the leaving group relative to the electron withdrawing ring substituent

11. Provide arrow-pushing mechanisms and the major product(s) expected for each of the following reactions:

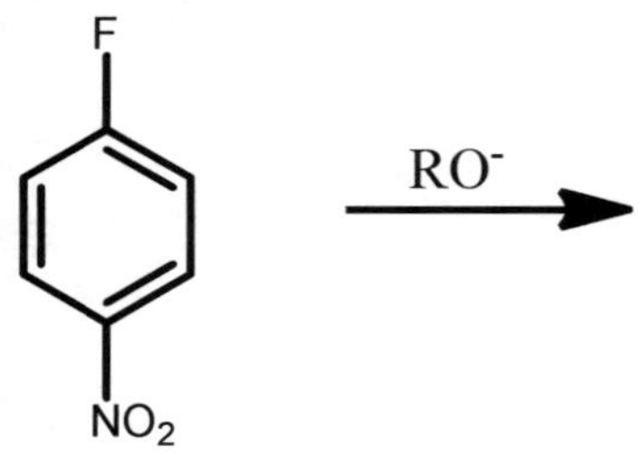

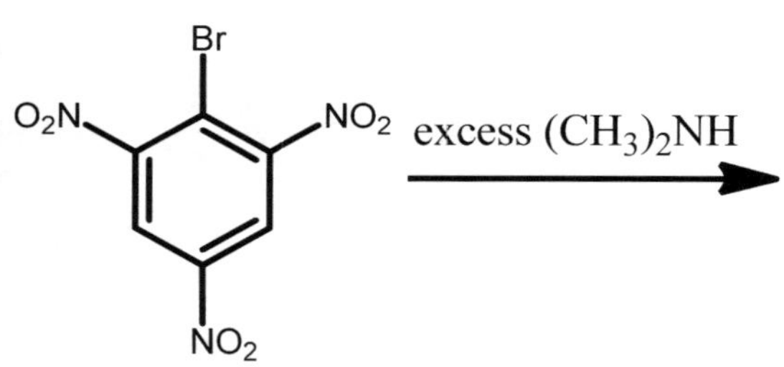

12. Strong bases can sometimes deprotonate benzene with concomitant elimination of a substituent such as a halogen, in a net step reminiscent of an E2 reaction. The resultant unstable intermediate can then undergo reaction with polar bonds or salts.

a) Draw the unstable intermediate and the final product of the following reaction sequence:

b) What would the product be if the E^+ Nu^- salt was PhMgBr?

13. The reactive intermediate **benzyne** can act as a dienophile for the Diels-Alder reaction. Provide an arrow-pushing mechanism for the following Diels-Alder reaction between a benzyne and a diene and draw the product in a manner that accurately represents the three-dimensional shape of the molecule.

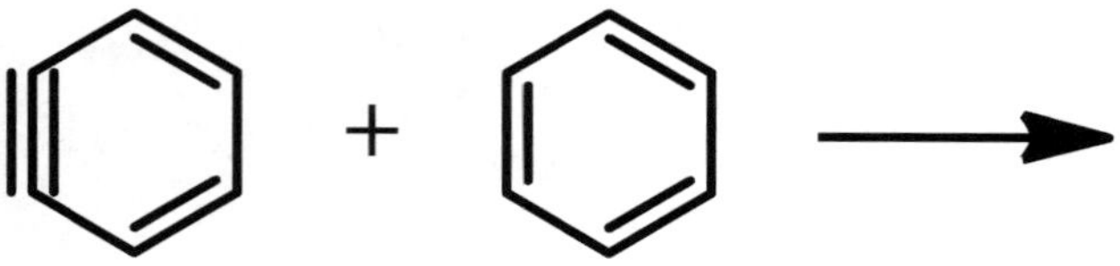

Problem Set 7: Nucleophilic Acyl Substitution

Date(s) covered in class:

Date by which material should be studied:

Chapter(s) in text covering material:

Date this problem set is due:

Points this problem set is worth:

1. Rank the relative reactivities of the carbonyl-containing functional groups discusses in class (more reactive first). Provide an explanation based on resonance and inductive effects for the rankings you provide.

2.

a) Provide a general mechanism for the nucleophilic acyl substitution reaction, using X and R as substituents on the carbonyl C and Nu^- as the nucleophile:

b) What properties of Nu and X dictate whether substitution will be favored in the forward reaction (irreversible), reverse reaction (reaction won't go to products) or will be reversible (some product and some reactant present)?

c) Provide a specific example of a nucleophilic acyl substitution reaction that would be reversible (near equilibrium between products and reactants).

d) Provide a specific example of a nucleophilic acyl substitution reaction that would be irreversible (products strongly favored).

3.

a) What property of carboxylic acids in particular makes them more difficult substrates for many nucleophilic acyl substitution reactions that, for example, acid chlorides?

b) Provide at least two examples of reactants that can be used to activate carboxylic acids in a way that alleviates the difficulties you discussed in part a).

c) give an arrow-pushing mechanism for the reaction of acetic acid with one of the two reagents that you chose in part b).

4. Provide the major product of this reaction, as well as a reasonable arrow-pushing mechanism for its formation.

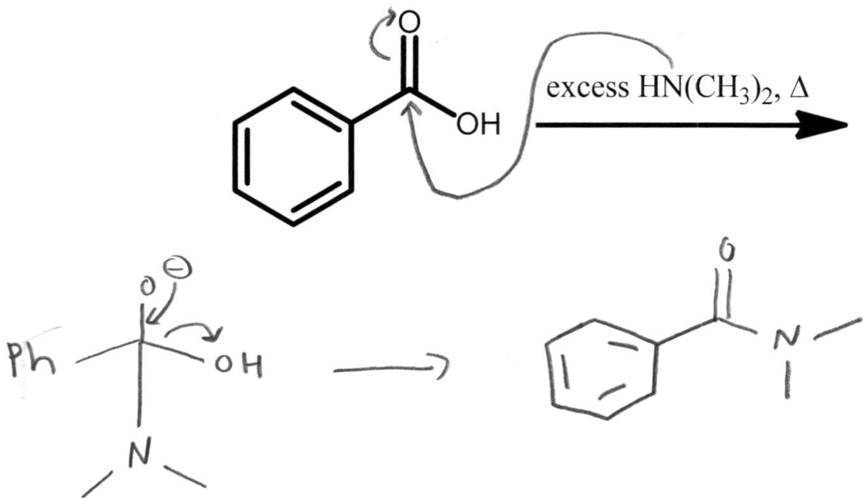

5. Predict the products of the following reactions:

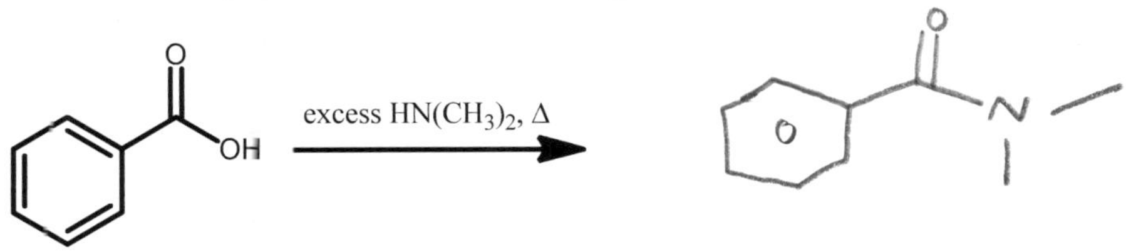

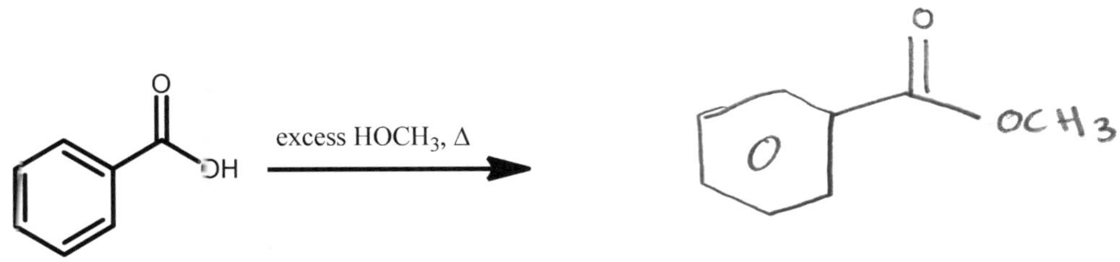

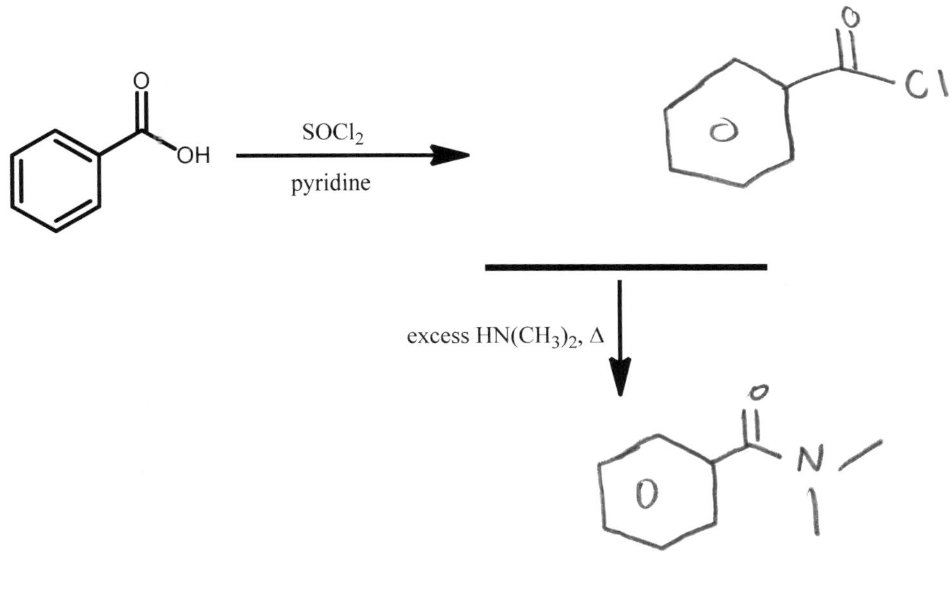

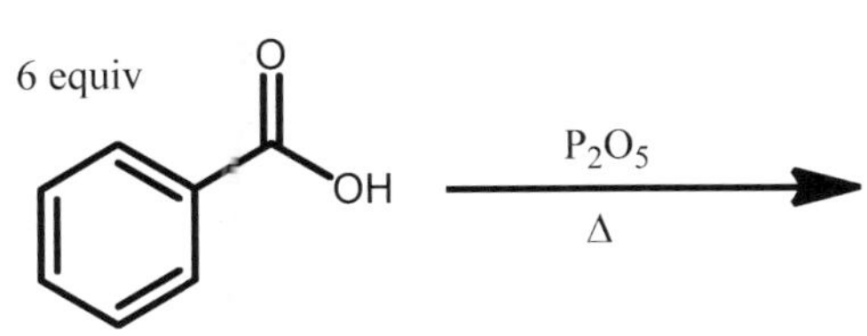

6. Provide the major product and a reasonable arrow-pushing mechanism for the base-catalyzed hydrolysis of ethyl acetate.

7. Provide the major product and a reasonable arrow-pushing mechanism for the acid-catalyzed hydrolysis of methyl benzoate.

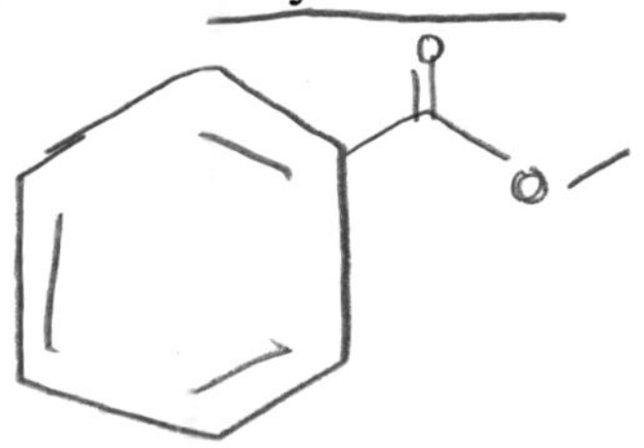

8. Predict the products of the following reactions:

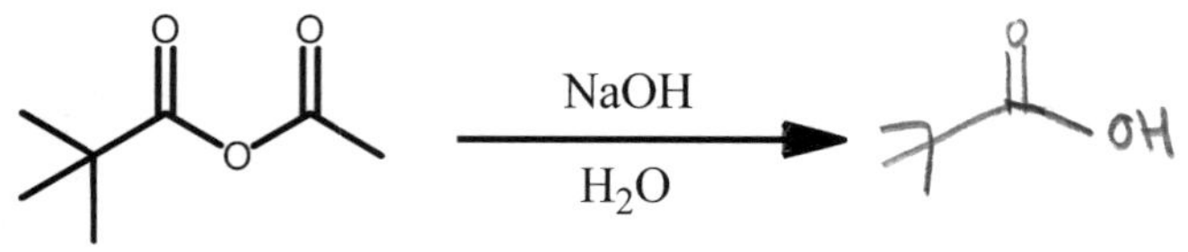

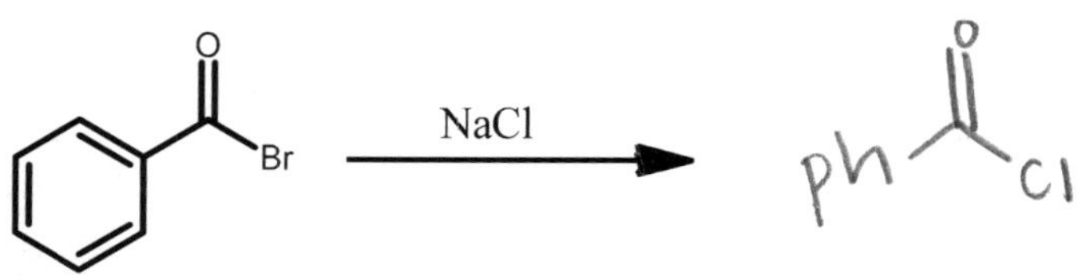

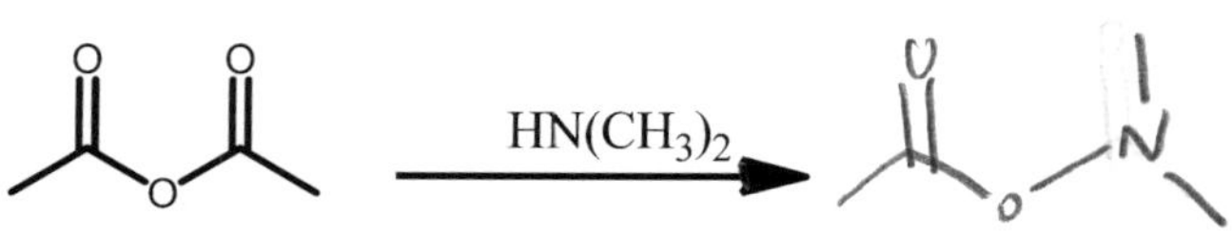

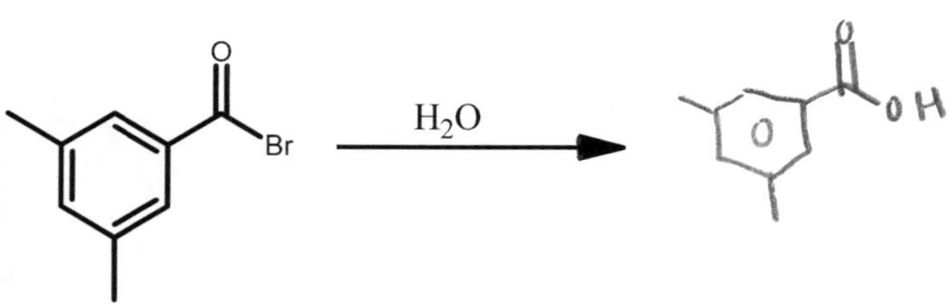

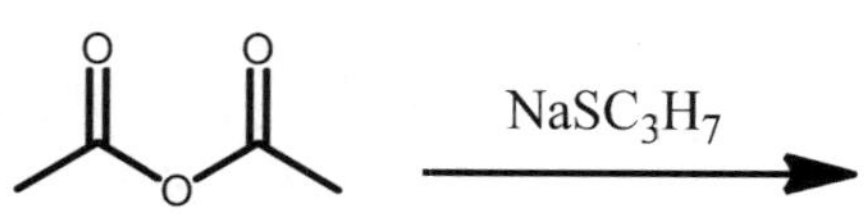

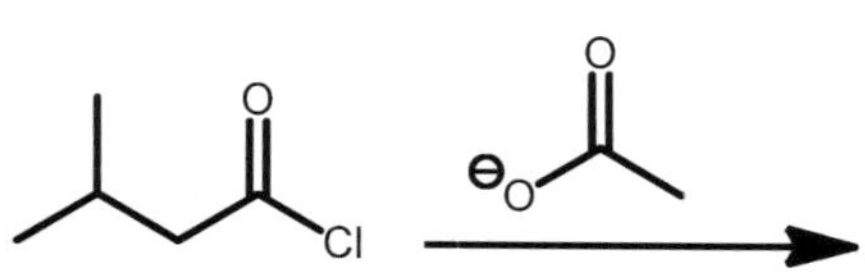

Problem Set 8: Nucleophilic Addition on Carbonyls

Date(s) covered in class:

Date by which material should be studied:

Chapter(s) in text covering material:

Date this problem set is due:

Points this problem set is worth:

1. a) Provide a general mechanism for nucleophilic addition to a carbonyl, using X and R as substituents on the carbonyl carbon and Nu^- as the nucleophile:

b) What properties of R and X dictate whether addition or substitution will be favored?

c) Provide a specific example of a nucleophilic addition to a carbonyl that would be irreversible and also strongly favored over substitution.

2. Predict the products in the blanks below. (One of these has no reaction!)

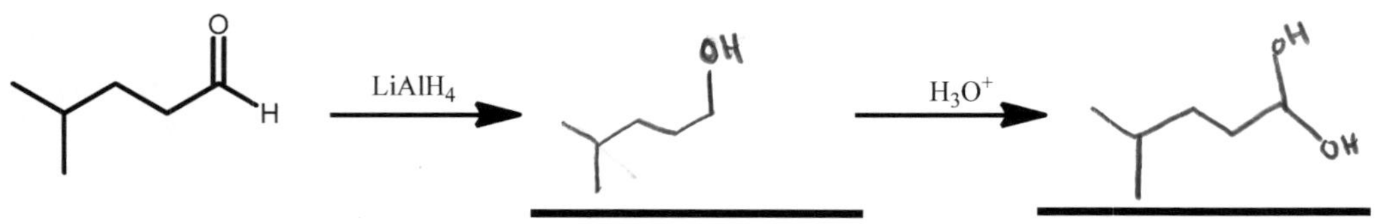

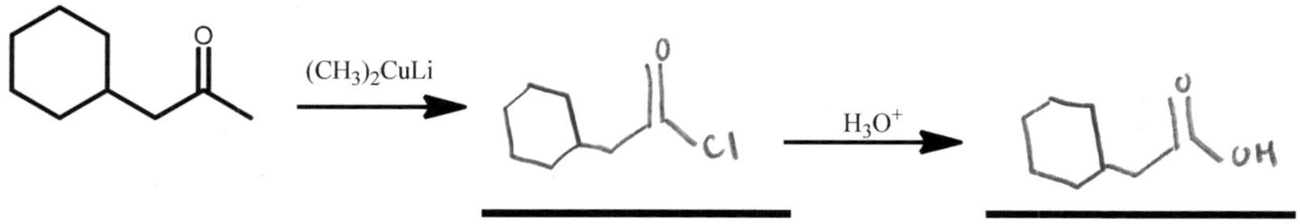

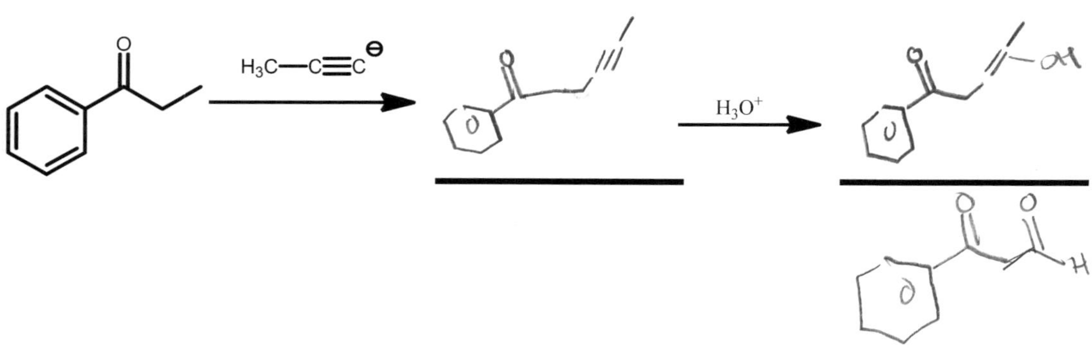

(cont'd)

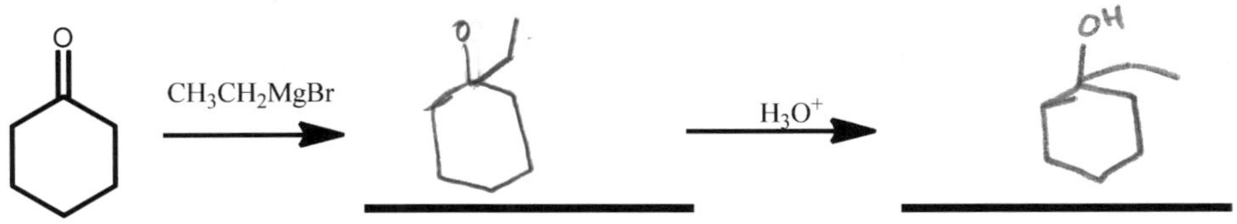
CH3CH2MgBr
H3O+

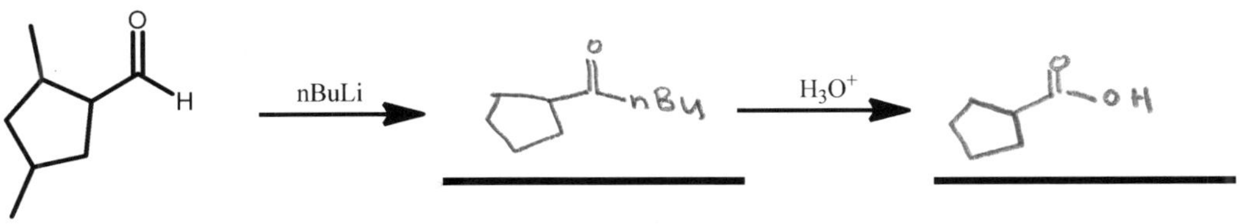
nBuLi
H3O+

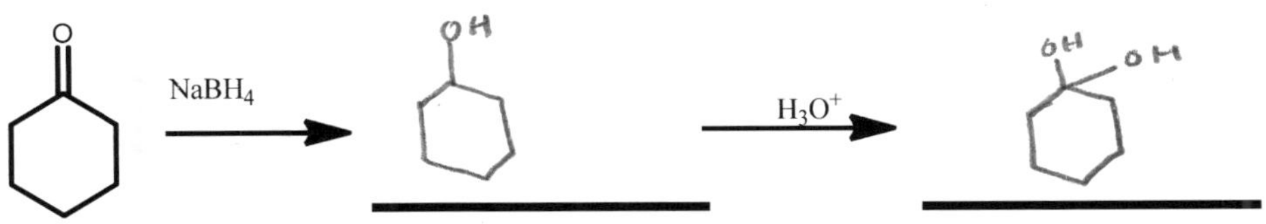
NaBH4
H3O+

3. Different hydride sources have different reactivities. Predict the products in the blanks below. Hint: each reagent will produce a different product.

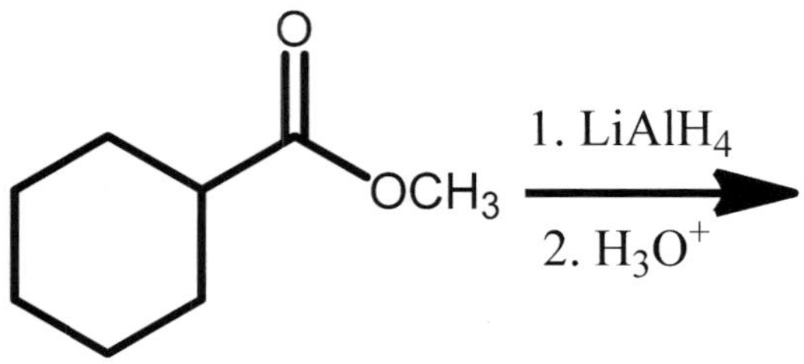

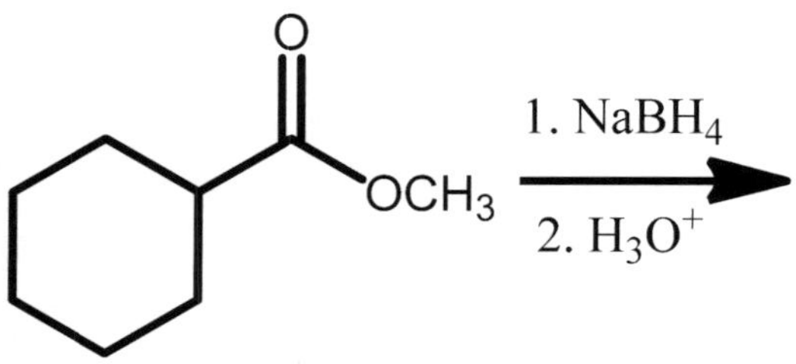

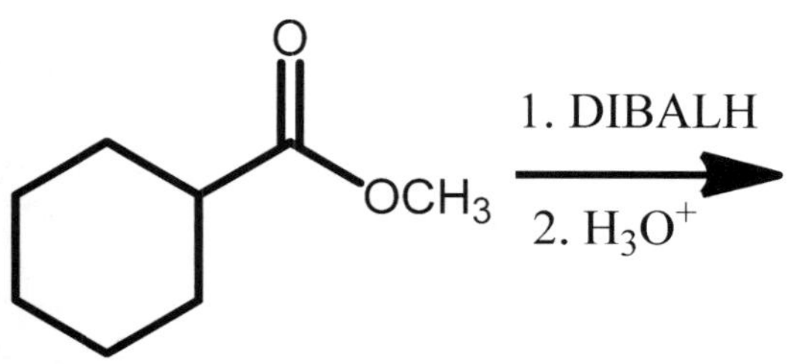

4. The various organometallic reagents (potential sources of R^-) have different reactivities. Predict the products in the blanks below.

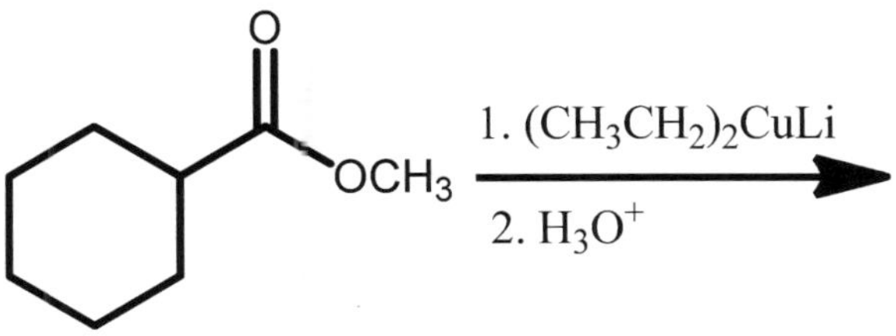

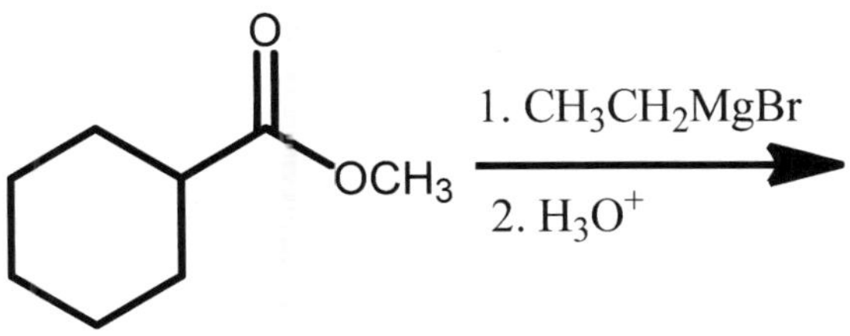

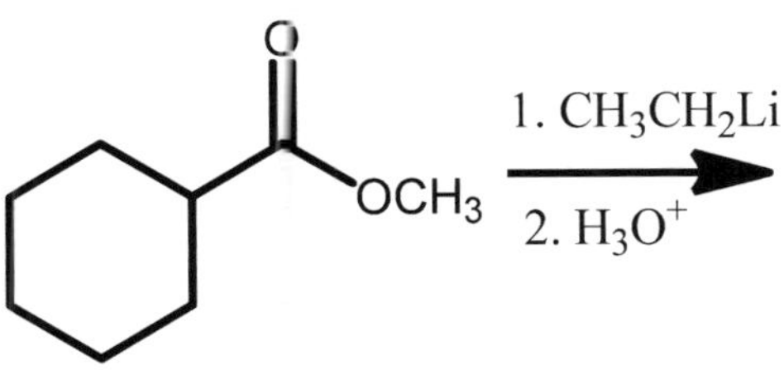

5. The different reactivities of organometallic reagents (potential sources of R^-) are especially noteworthy in how they react with α,β-unsaturated carbonyls. Provide the major product for each of the following reactions:

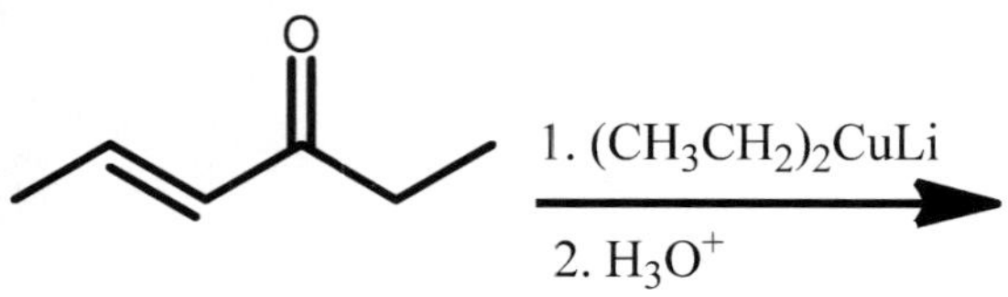

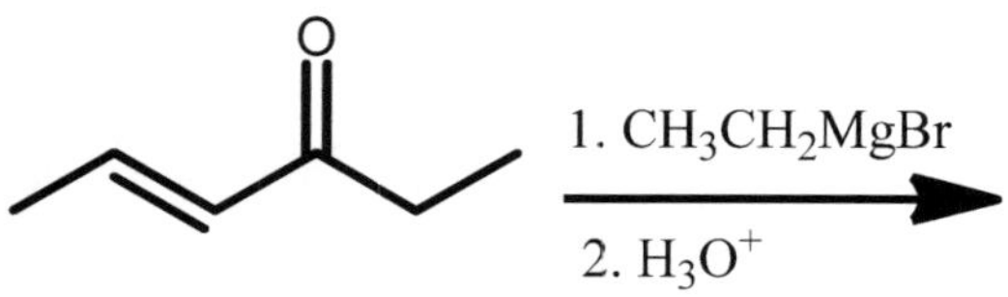

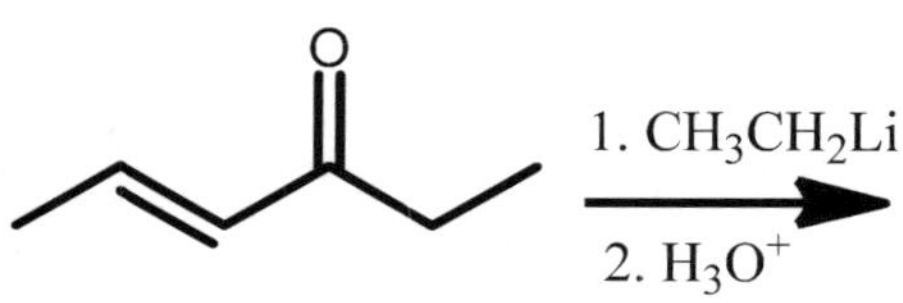

6. Predict the product for each of the following reactions of the acid chloride shown with each of the following reaction conditions:

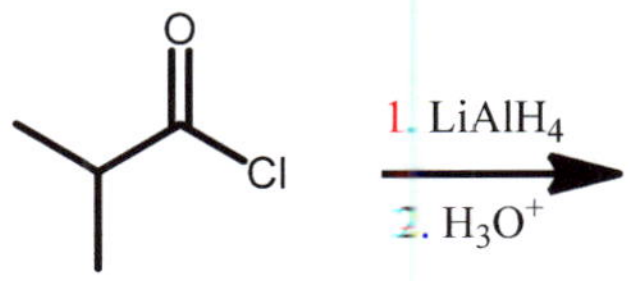

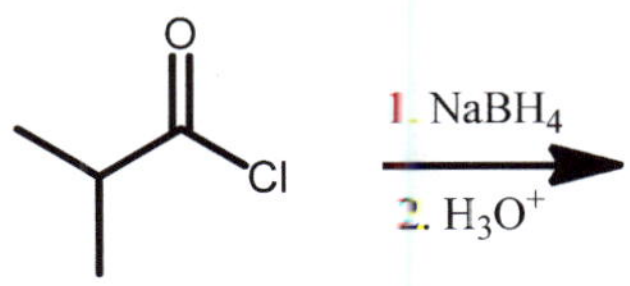

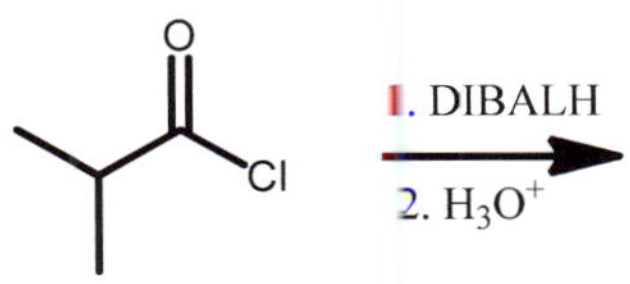

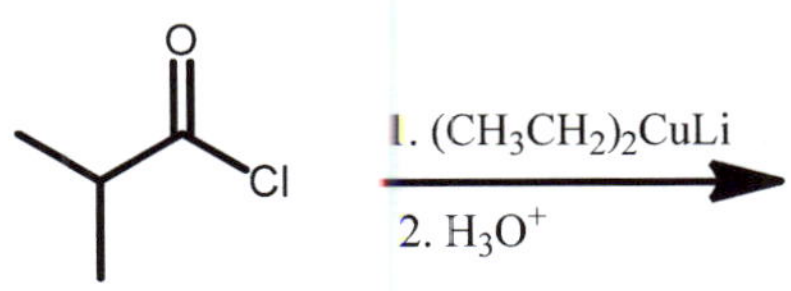

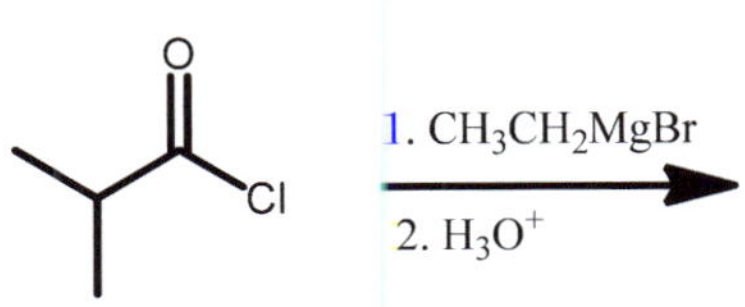

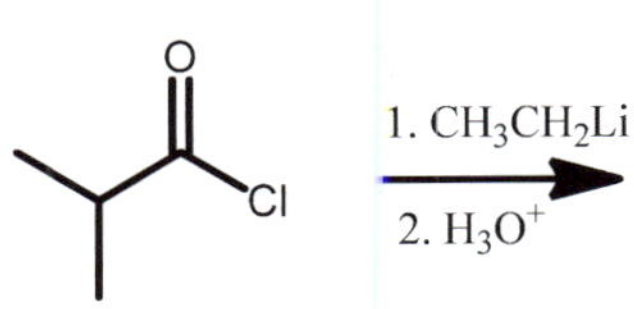

7. Provide a reasonable arrow-pushing mechanism and major product of the following reaction.

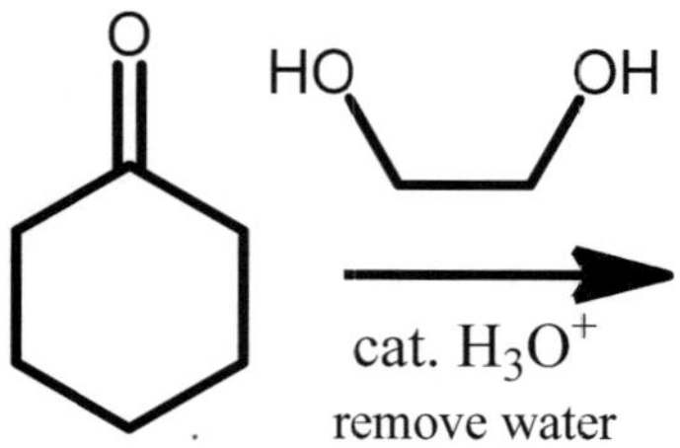

8. Provide the major product for each of the following reactions. Keep the difference in reactivity of the two different carbonyl functional groups in mind.

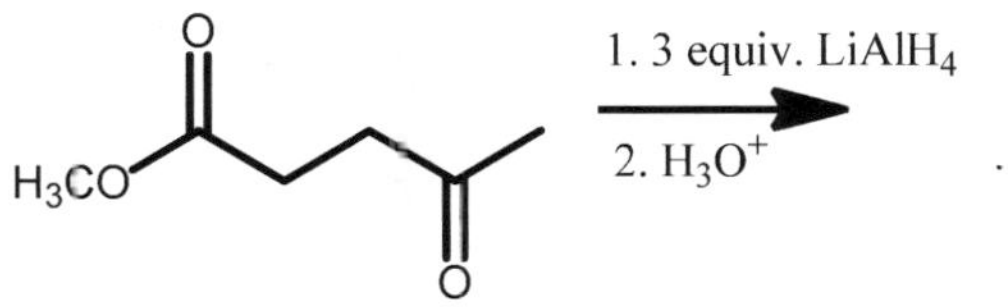

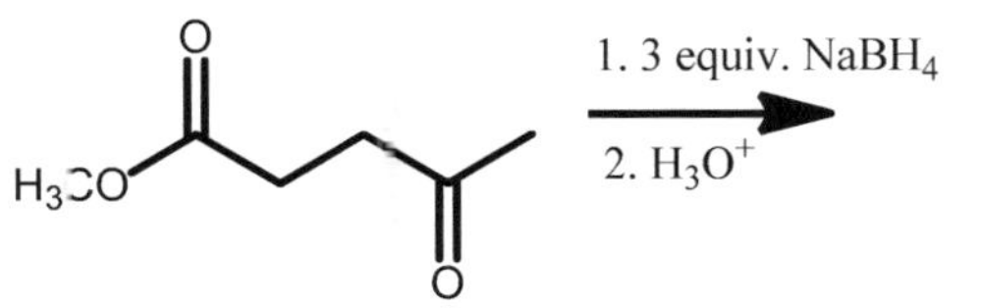

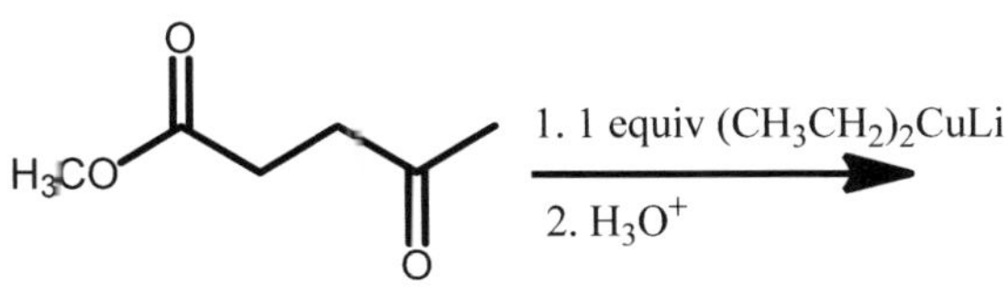

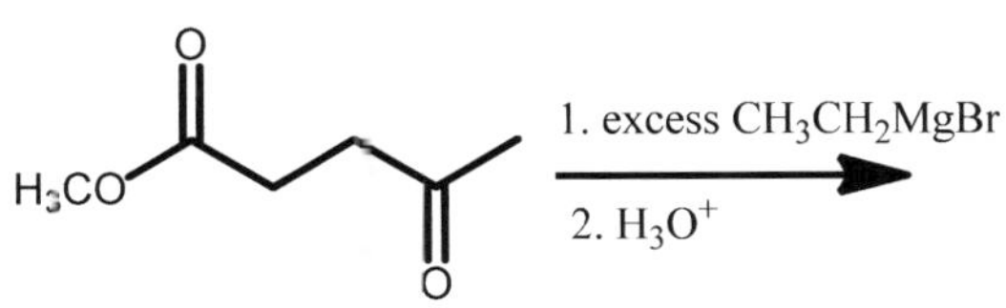

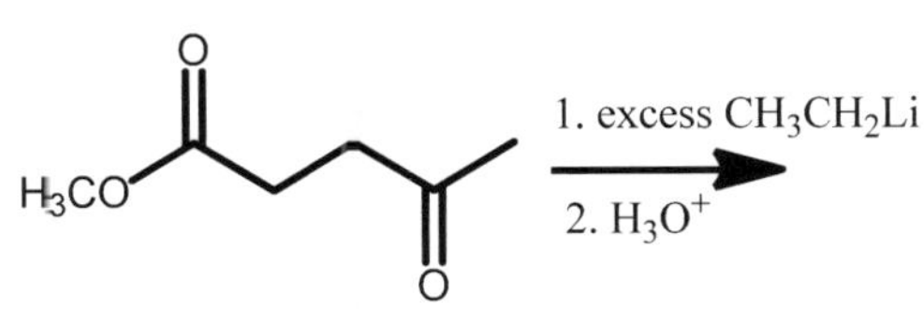

9. Provide a reasonable arrow-pushing mechanism for the reaction of the following aldehyde and ylide, and clearly label the major product. Be sure to draw the product in the correct configuration.

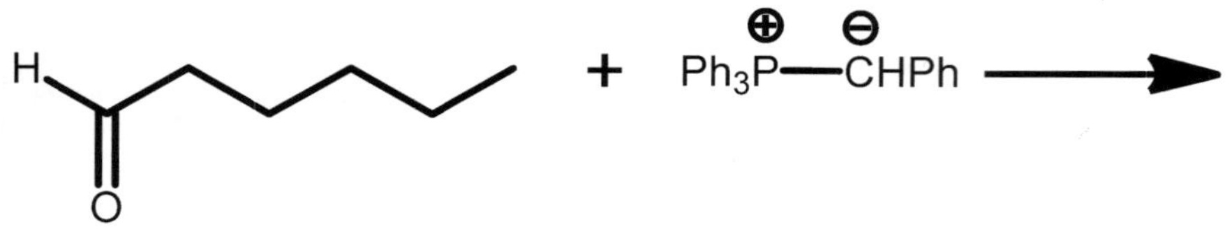

10. Deuterium labeling is an excellent tool for probing reaction mechanisms. For each of the following reactions involving deuterated reagents, provide a reasonable arrow-pushing mechanism and predict the major products. Be sure to show ALL of the major products (for example, if ROH or NaCl is produced, show those in addition to the highest molecular weight organic product).

a)

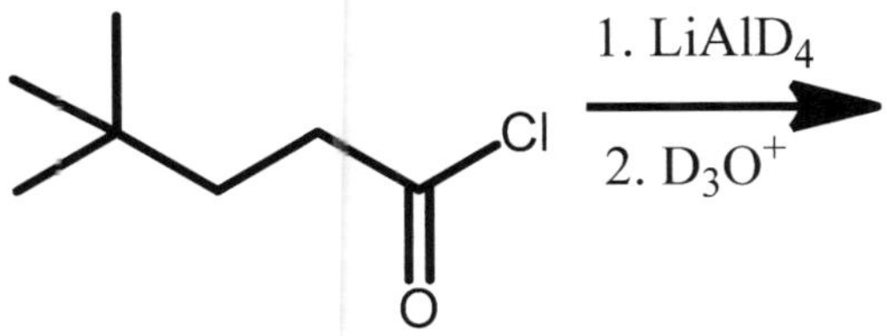

b)

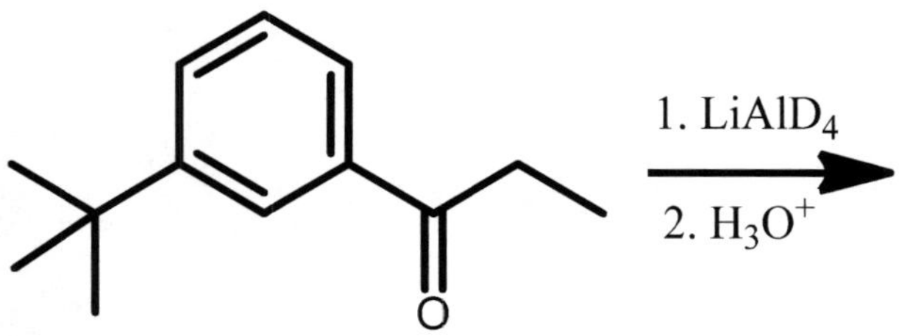

c)

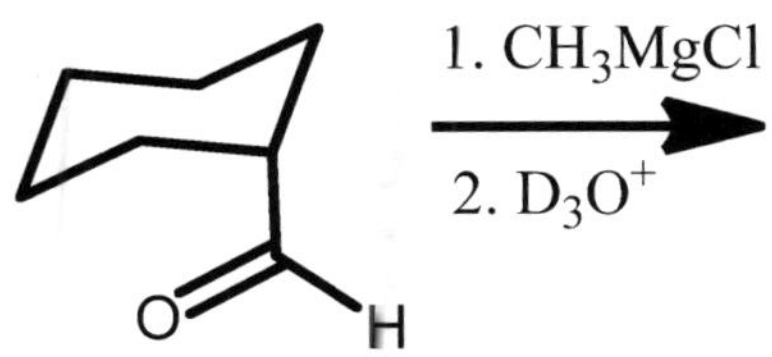

d)

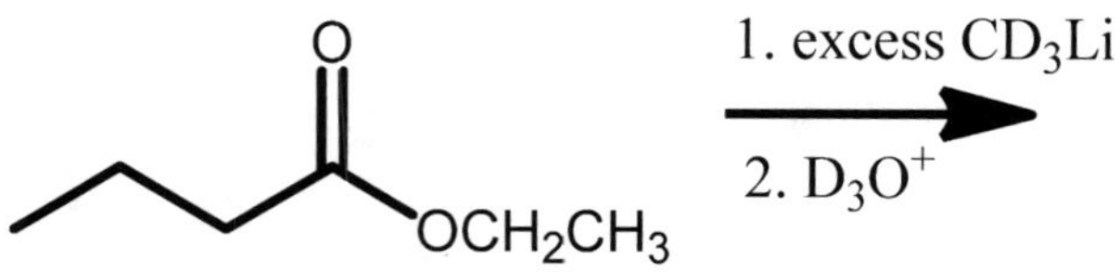

11. Provide the reagents and a reasonable arrow-pushing mechanism showing how each of the following could best be produced from the provided starting materials.

a)

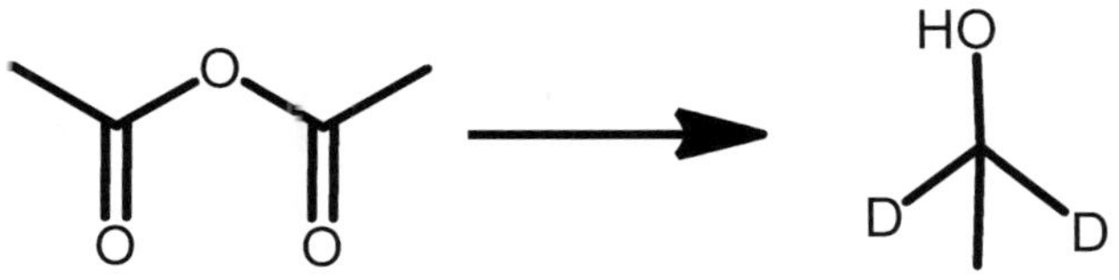

b)

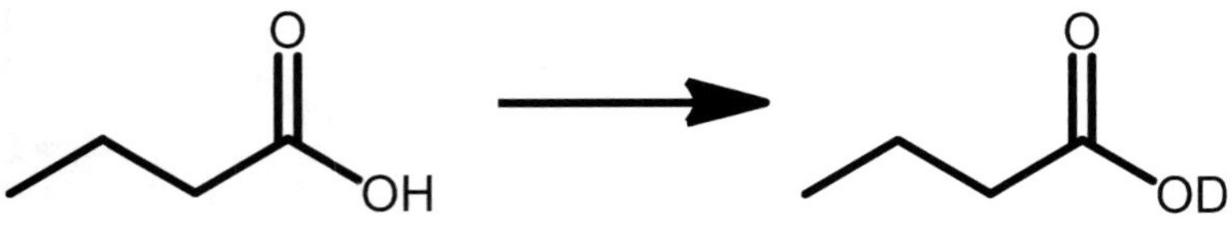

c)

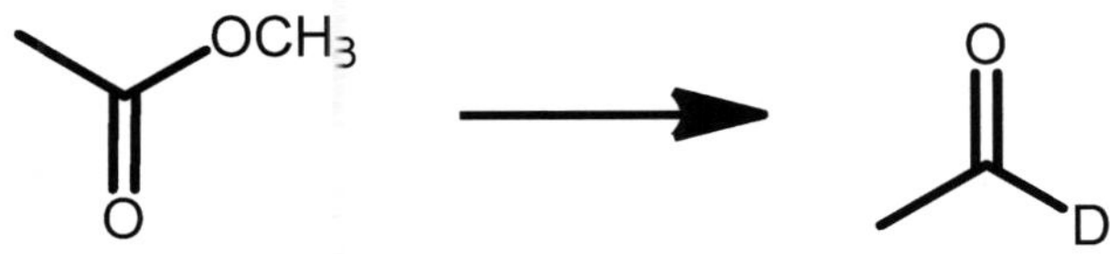

Problem Set 9: Reactions involving the Carbonyl α-Carbon

Date(s) covered in class:

Date by which material should be studied:

Chapter(s) in text covering material:

Date this problem set is due:

Points this problem set is worth:

1.

a) Provide the structure of an enol and an enolate

b) Provide an example of a carbonyl that has a proton on the alpha carbon that is more easily removed by a base than is the alpha proton of acetone

c) Provide an example of an enol that is more stable than its keto tautomer. Why is the enol form more stable in the example that you chose?

2. Circle the molecules that **cannot** form an enolate via reaction with a base.

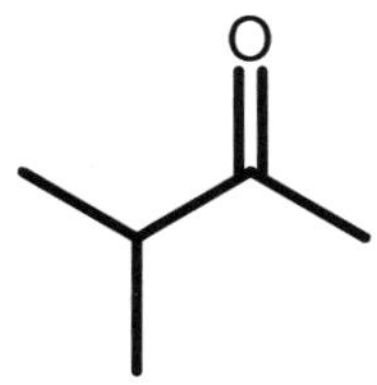

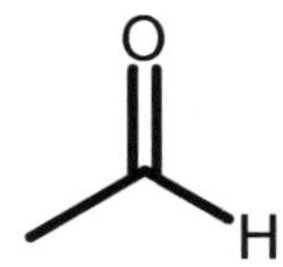

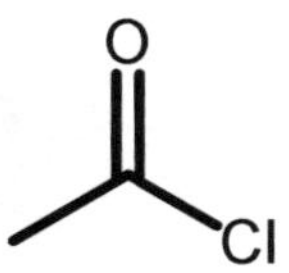

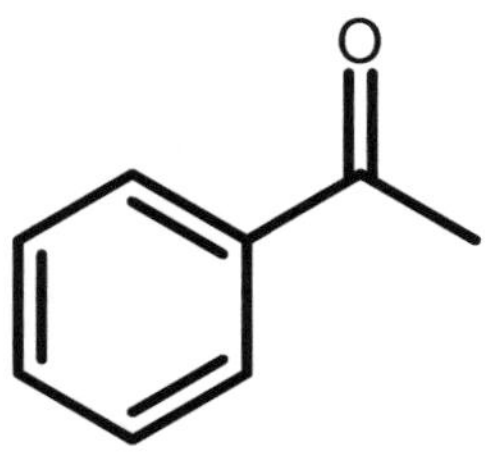

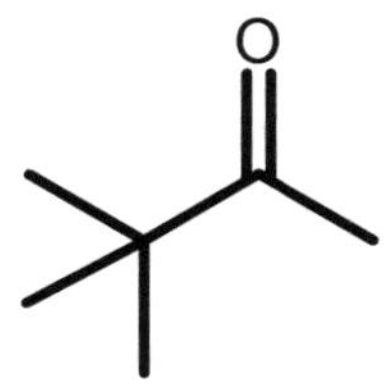

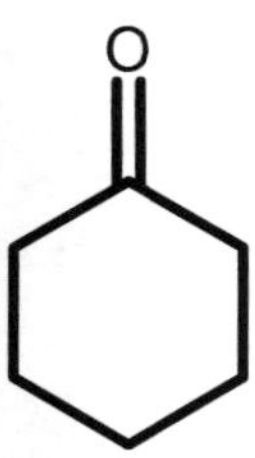

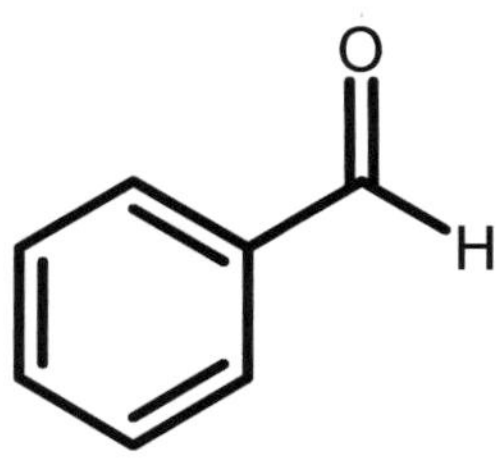

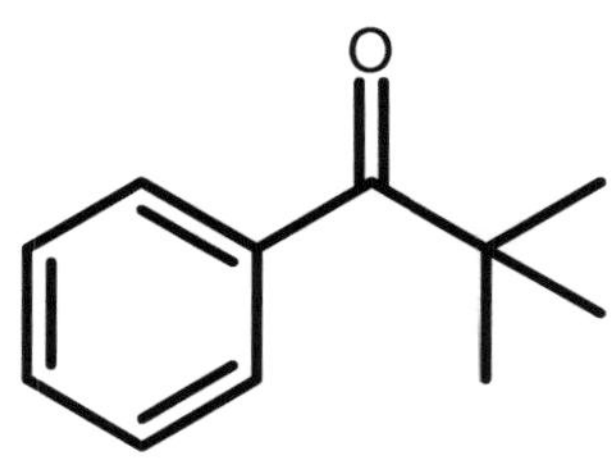

3. Provide arrow-pushing mechanisms for the following reactions that employ an enolate as the nucleophile to attack another carbonyl-bearing molecule. Where necessary, provide any acid or base needed to produce a neutral compound as the final product. For convenience, all enolates are drawn in the resonance contributor with the negative charge on the carbon atom.

3a)

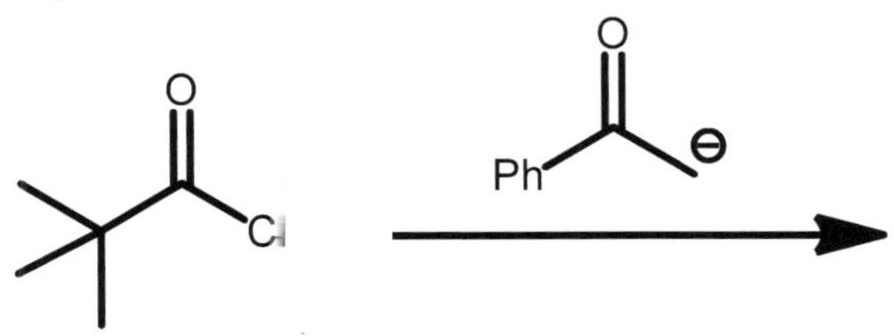

3b)

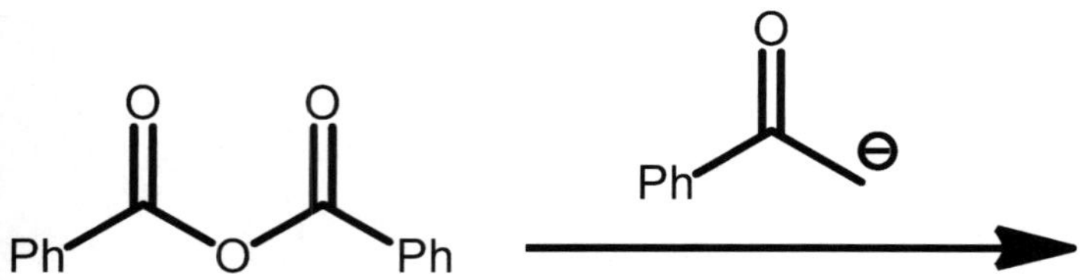

3c)

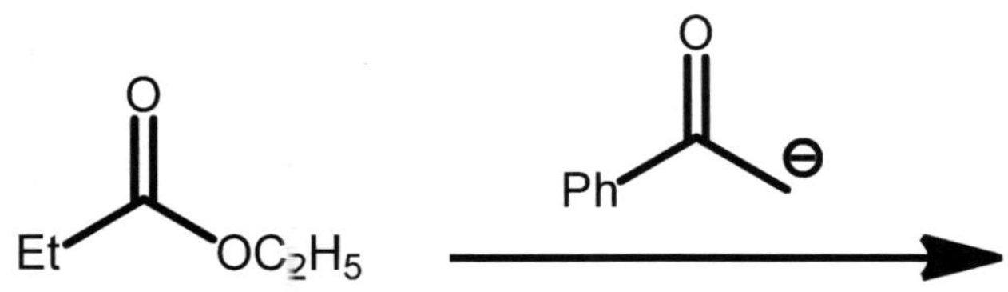

3d)

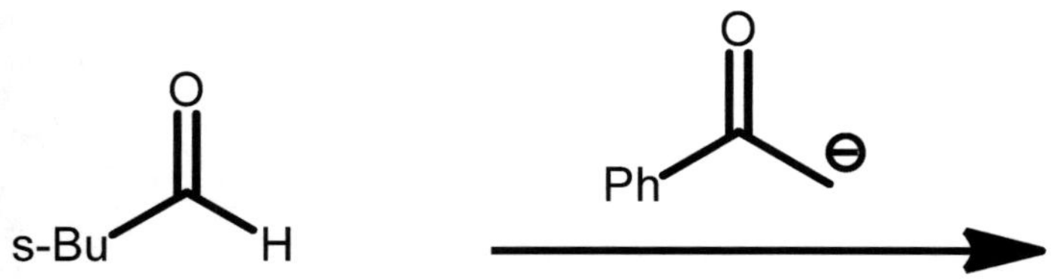

3e)

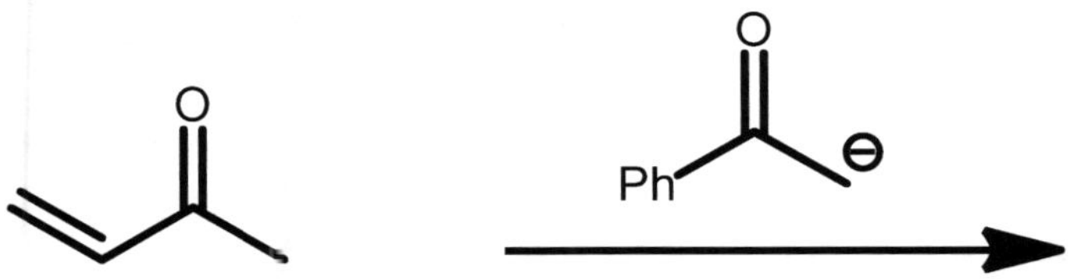

3f)

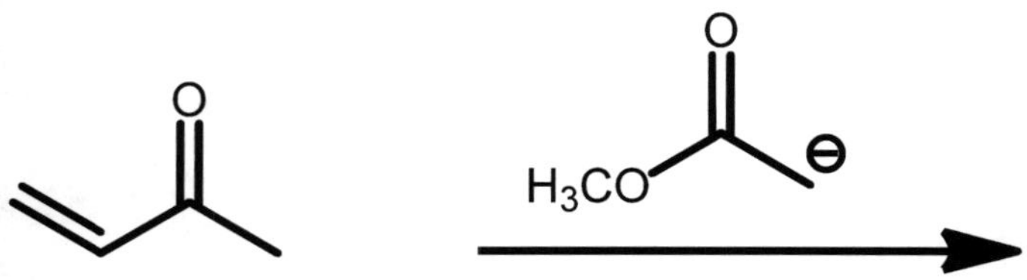

3g)

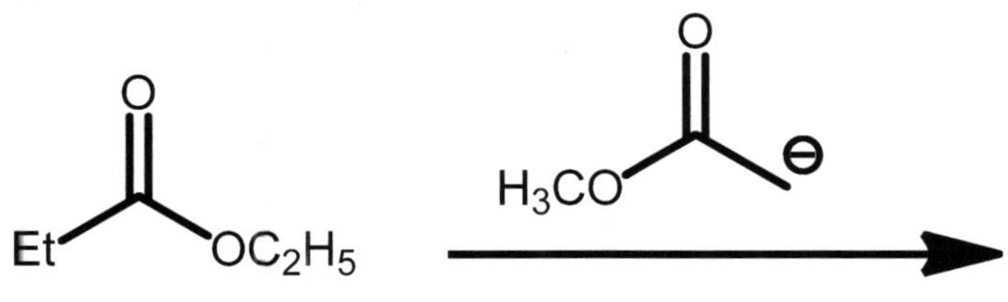

4. a) Which is most easily deprotonated at the alpha position, acetone or ethyl acetate? Explain why.

b) What will be the major organic product if a base is added to a 1:1 mixture of acetone and ethyl acetate?

5. Draw **all** possible Aldol addition products resulting from the reaction below, and circle the major product. Provide a rationale for why the product selected would be the major product.

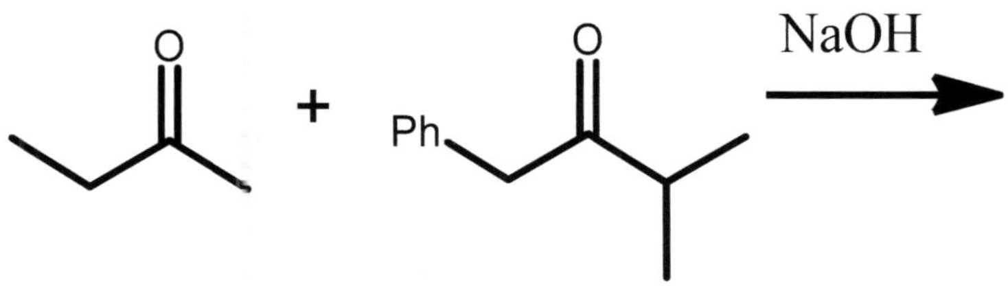

6. Draw all of the possible Aldol condensation products that could form if the previous reaction was heated.

7. Provide arrow-pushing mechanisms for the following reaction, and clearly indicate the major product.

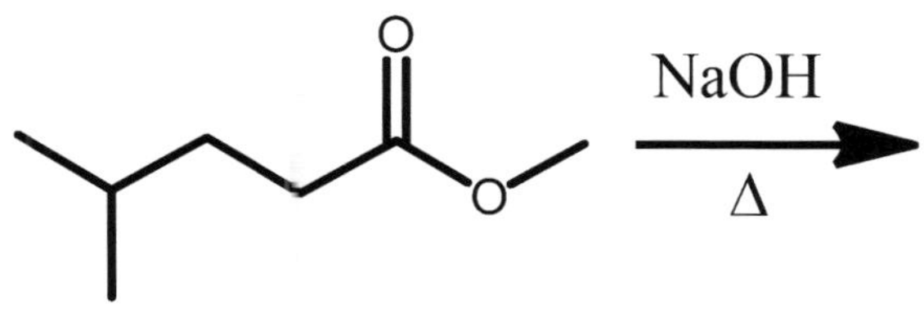

8. Provide arrow-pushing mechanisms for the following reaction, and clearly indicate the major product.

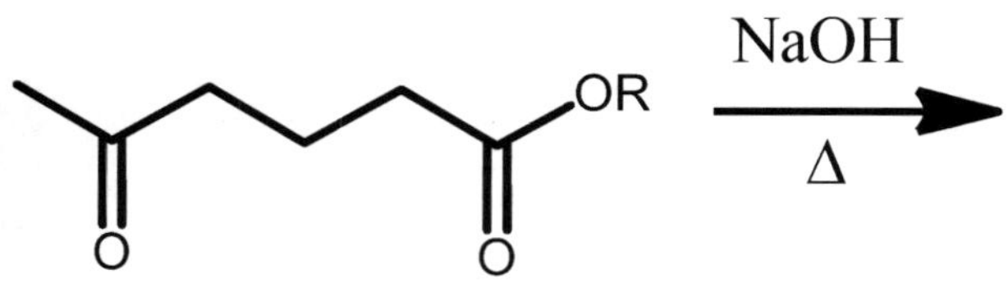

9. Provide arrow-pushing mechanisms for the following reaction, and clearly indicate the major product.

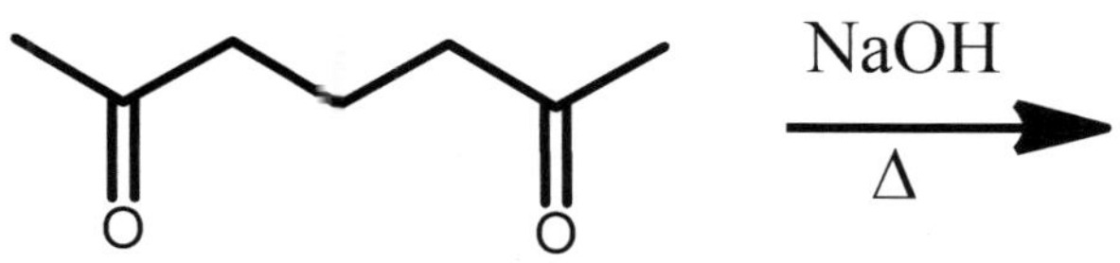

10. Provide an arrow-pushing mechanism showing the halogenations of acetone with bromine.

11. Provide an arrow-pushing mechanism for an example of the haloform reaction. Select an example that was not discussed in lecture and that produces bromoform as one of the products.

12. Provide an arrow-pushing mechanism for the alkylation of 3-pentanone, employing 2-bromopentane as a reagent.

13. Provide an arrow-pushing mechanism for the decarboxylation of hexanoic acid.

14. Provide examples (not provided in lecture or your text) of each of the following reactions (no mechanisms required).

a) Dieckmann Condensation

b) Claisen Condensation

c) Acetoacetic Ester Synthesis

d) Malonic Ester Synthesis

e) Robinson Annulation

f) Michael Addition

Problem Set 10: Reactions of Carbonyls with Amines / Amine Derivatives; Reactions of Nitriles and Imines

Date(s) covered in class:

Date by which material should be studied:

Chapter(s) in text covering material:

Date this problem set is due:

Points this problem set is worth:

1. Provide an arrow-pushing mechanism for the following reaction, clearly labeling the major final product.

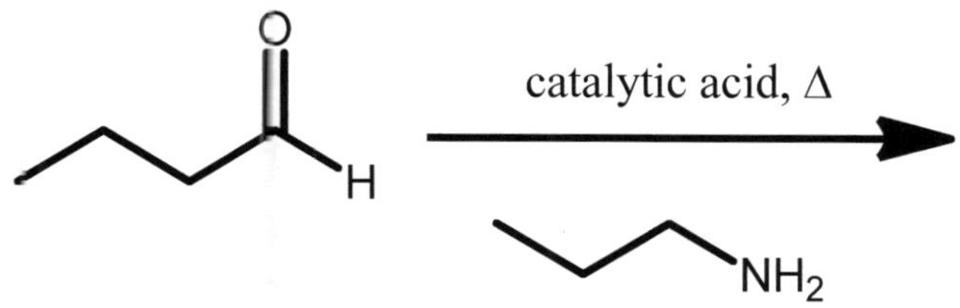

2. Provide the major product for the following reactions:

a)

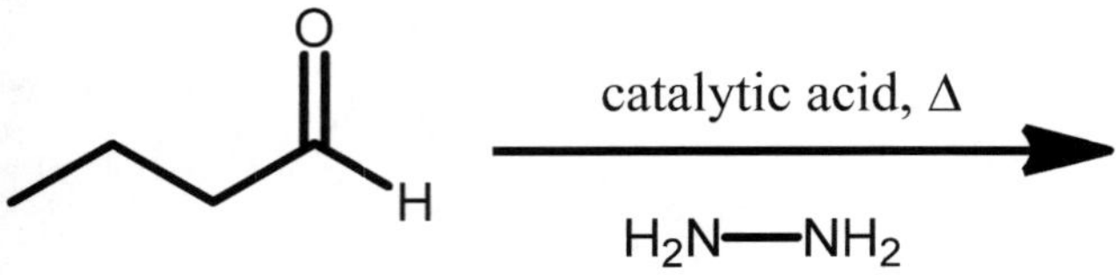

b)

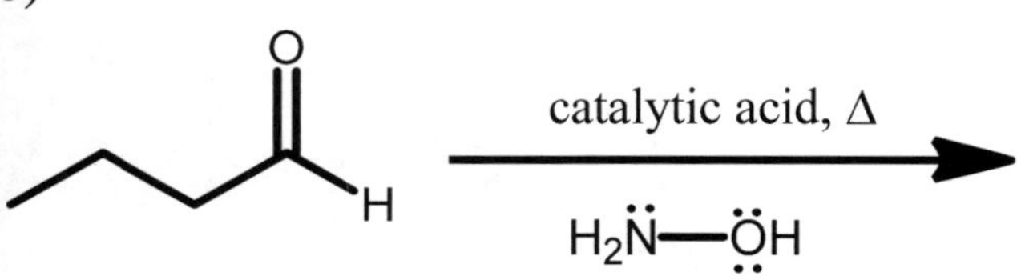

3. In the following reaction, there are three nitrogen atoms with lone pair electrons that one might consider as nucleophilic positions. Indicate which nitrogen is the most nucleophilic. Explain your selection on the basis of the resonance contributors to the compound's electronic structure.

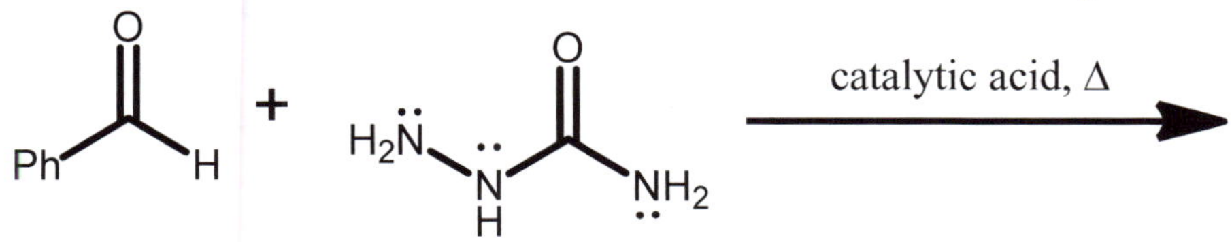

4. Provide an arrow-pushing mechanism for the following reaction, clearly labeling the major final product.

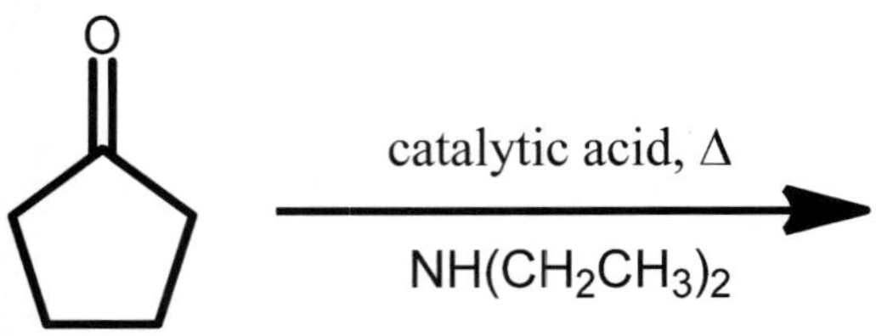

5. Provide an arrow-pushing mechanism for the following reaction, clearly labeling the major final product.

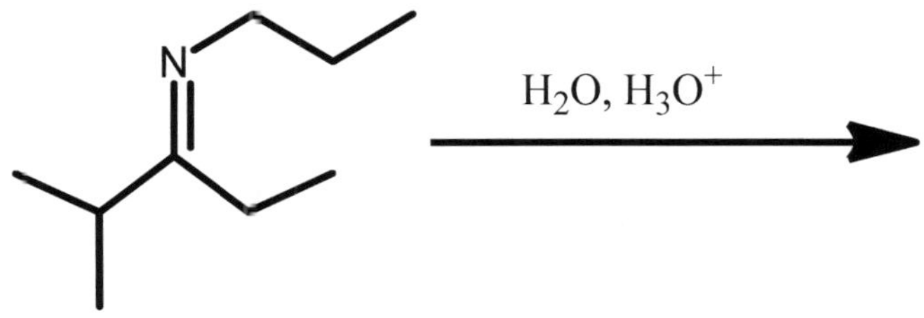

6. Provide the final major product for each of the following reactions.

a)

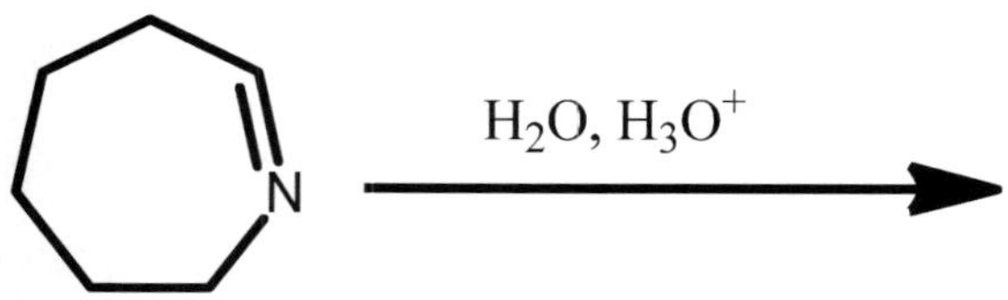

b)

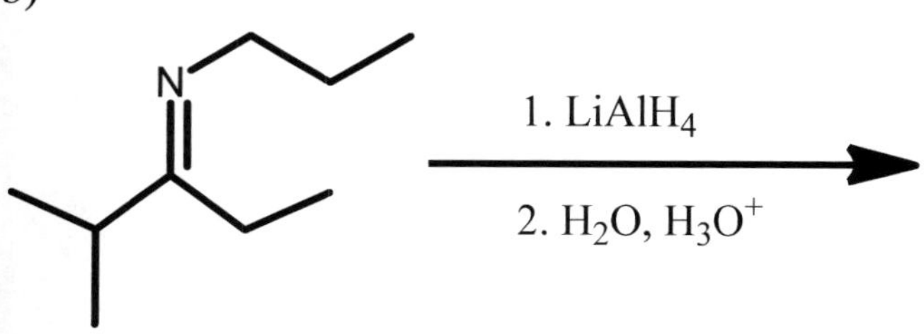

7. Provide the structures of compounds A, B, C and D.

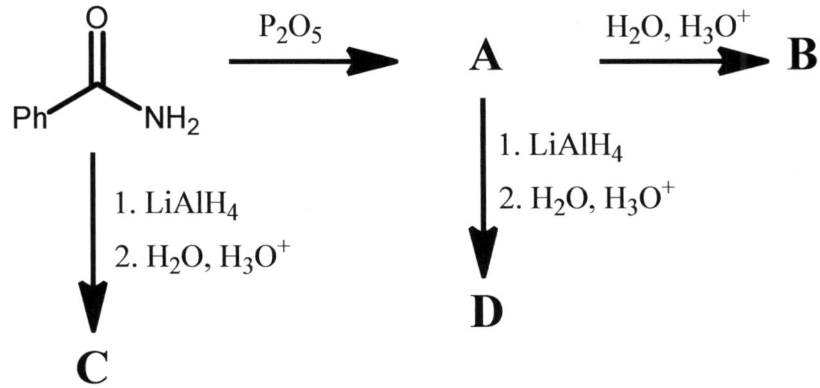

Progress Check 1:

Material Coverage:

"Organic Chemistry 2 Lecture Guide 2013":

Before We Begin and Lecture Sets 1, 4 and 5

"First Semester Organic Chemistry Reactions 2012: Everything in One Place":

Sections I, II.1-11.2, II, IV.

Date by which material should be studied:

Chapter(s) in text covering material:

Date this problem set is due:

Points this problem set is worth:

PROGRESS CHECK 1

1. What are the hybridizations of the atoms marked **A**, **B**, and **C** in the molecule?

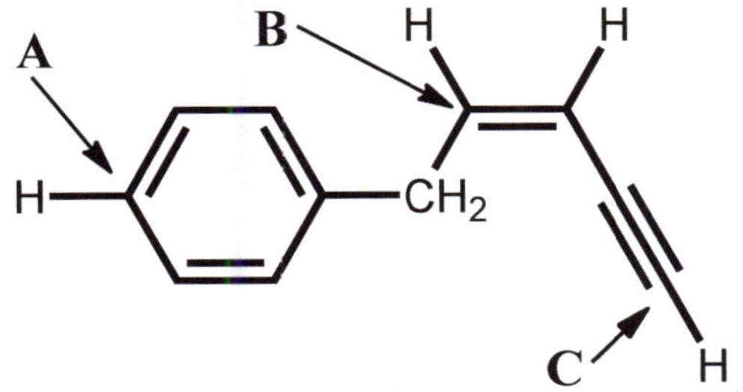

(A) A = *sp* **B** = *sp* **C** = sp^2
(B) A = sp^2 **B** = sp^2 **C** = *sp*
(C) A = sp^2 **B** = sp^2 **C** = sp^3
(D) A = sp^2 **B** = sp^3 **C** = *sp*

2. Which is the most stable conformation of *cis*-1-ethyl-2-isopropylcyclohexane?

(A) 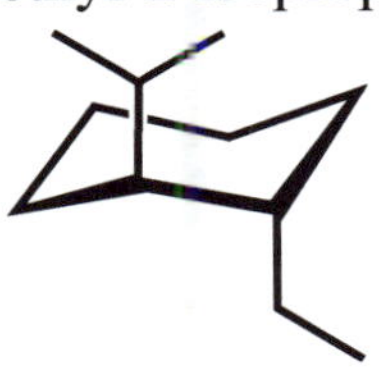

(B) 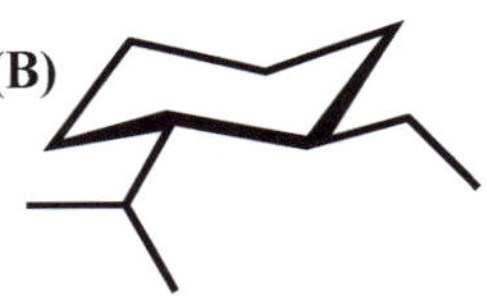

(C) 

(D) 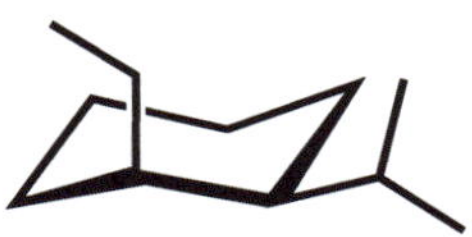

3. Which is the ***weakest*** acid?

(A) 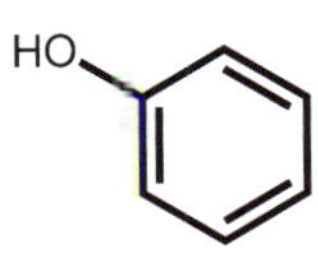

(B) 

(C) 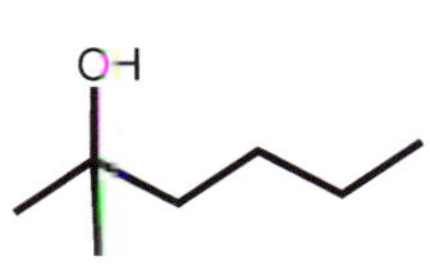

(D) 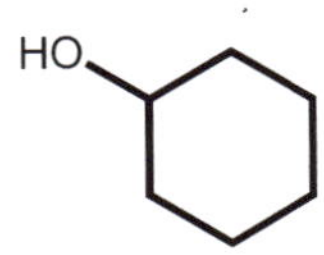

4. What is a major product of this reaction?

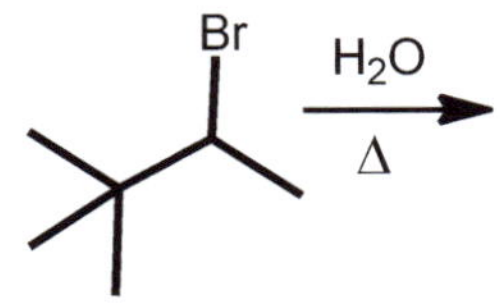

(A) 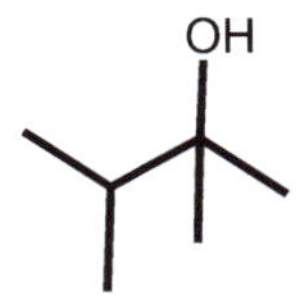

(B) 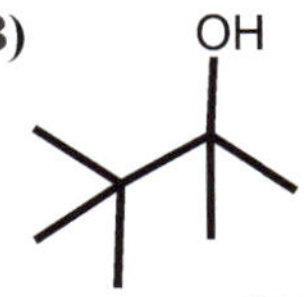

(C) 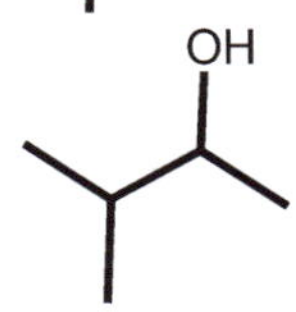

(D) 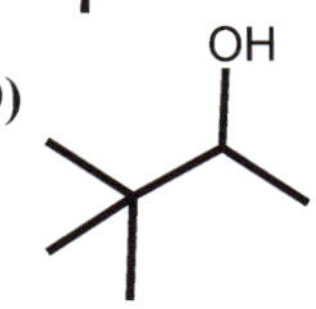

5. Which reagent is needed to produce the indicated product via a **Suzuki** coupling reaction?

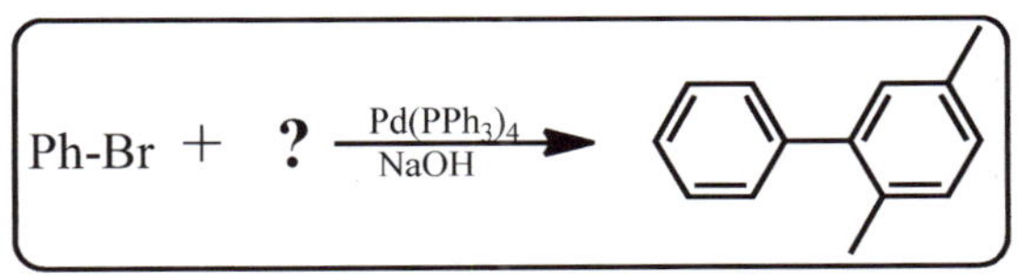

(A) 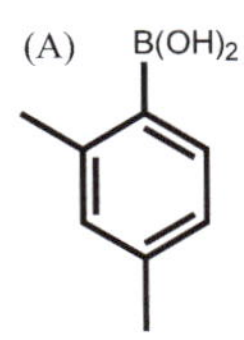

(B) 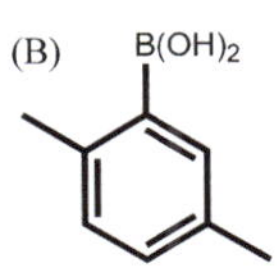

(C) 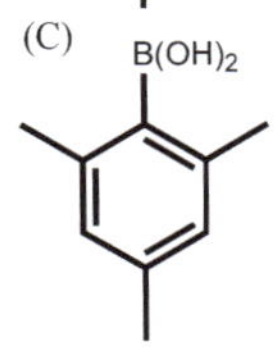

(D) 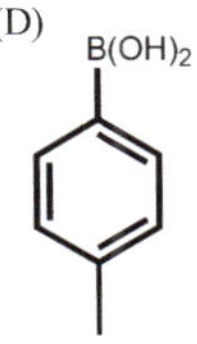

6. What is the product of the Diels-Alder reaction shown?

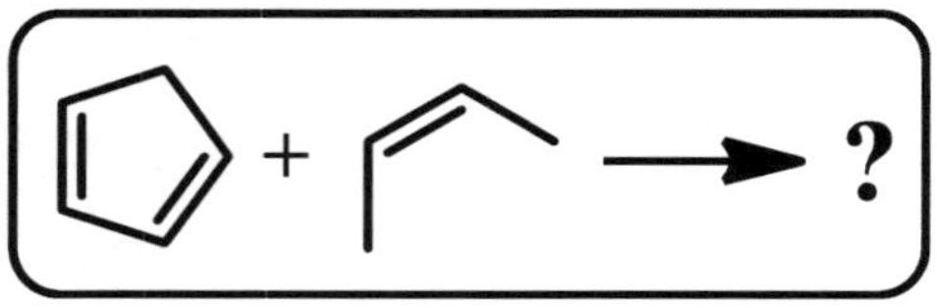

(A)  (B) 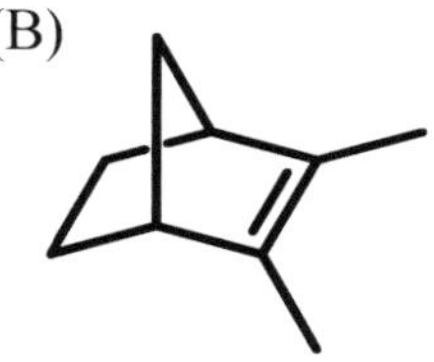

(C)  (D) 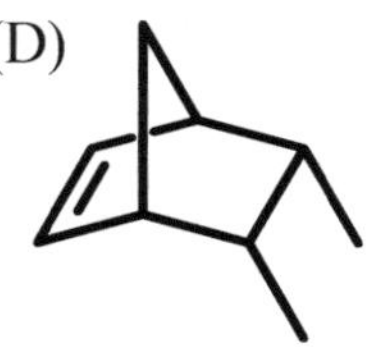

7. When 1-iodobutane is heated in the presence of water, the major product is:

(A) 1-butanol only
(B) 2-butanol only
(C) a mixture of 2-butanol and *trans*-2-butene
(D) No reaction

8. When (*S*)-2-iodobutane reacts with excess water upon heating, the major substitution product is:

(A) (*S*)-2-butanol
(B) (*R*)-2-butanol
(C) A racemic mixture of (*R*)- and (*S*)-2-butanol
(D) No reaction

9. What is the major product of the reaction sequence?

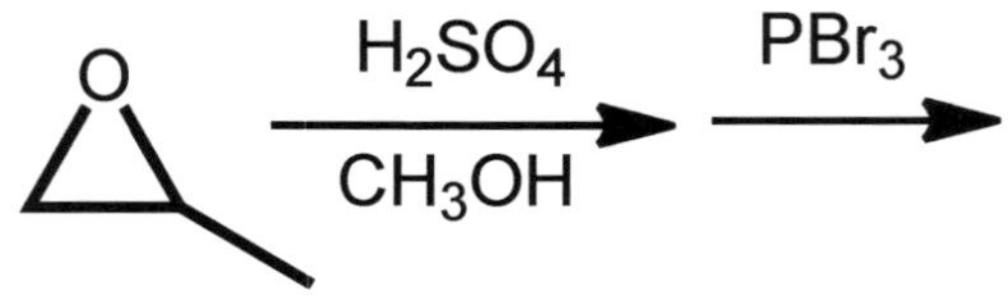

(A) 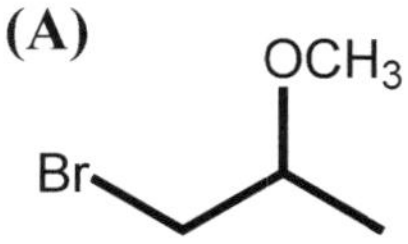

(B) 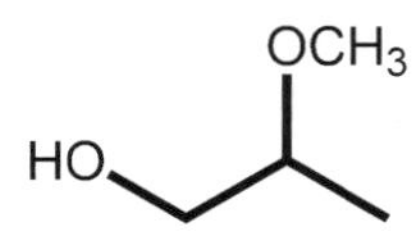

(C) 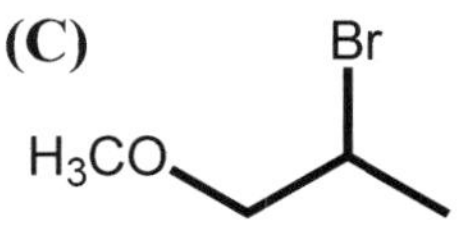

(D) 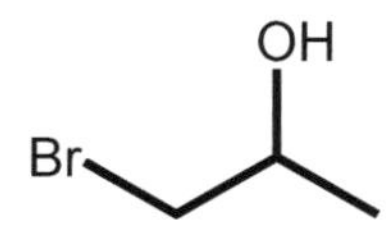

10. Which reaction series has a bromonium ion intermediate in the mechanistic steps leading to the major product?

(A) 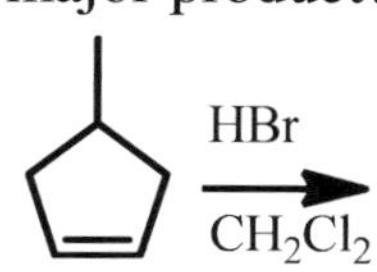

(B) 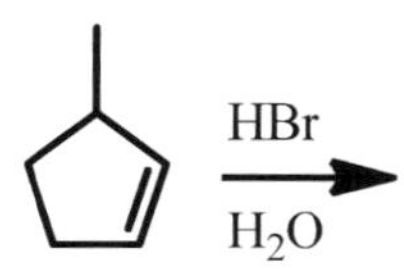

(C) 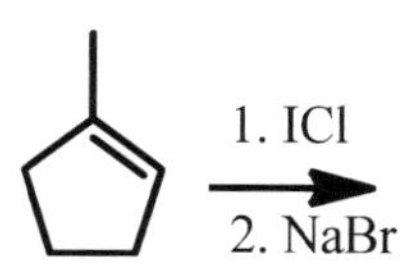

(D) 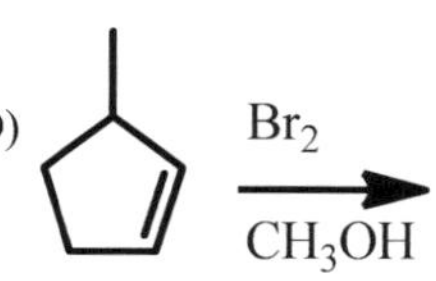

11. Which anion is aromatic?

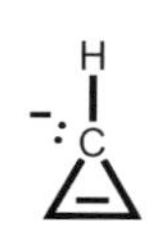
(A)
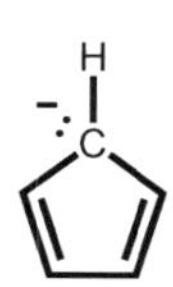
(B)
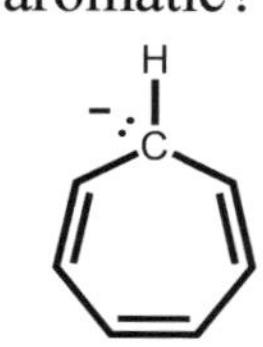
(C)

(D) All are aromatic

12. Which cation is **not** aromatic?

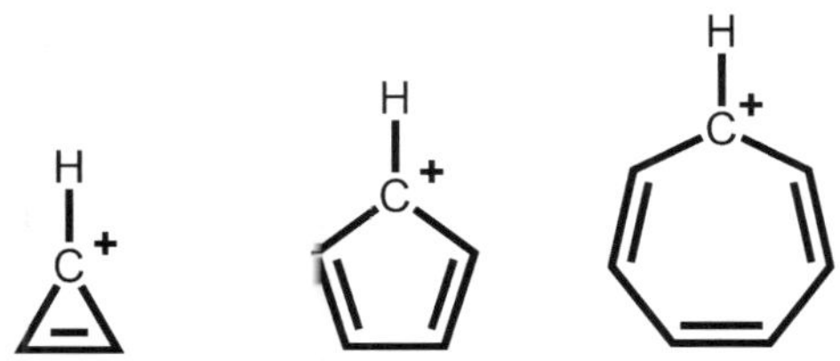

(A) (B) (C)

(D) All are aromatic

13. Which is the most basic compound?

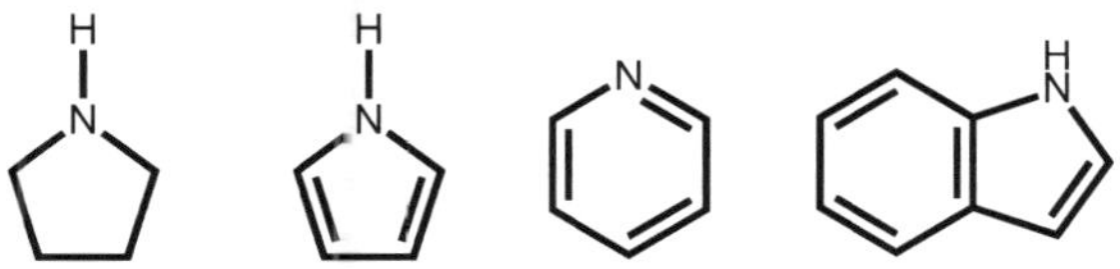

(A) (B) (C) (D)

14. Which reacts fastest by an **electrophilic** aromatic substitution reaction?

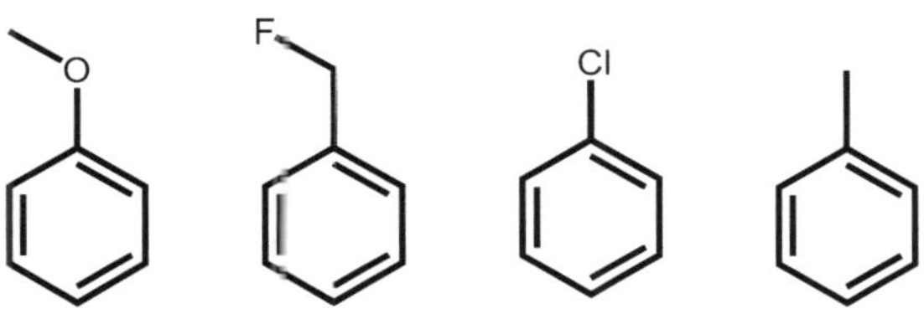

(A) (B) (C) (D)

15. Which of the following best describes the tosyl group?

A) A strong base
B) A good leaving group
C) A good nucleophile
D) A strong acid

16. Which **site** will react fastest to nitration (or any electrophilic aromatic substitution) reaction?

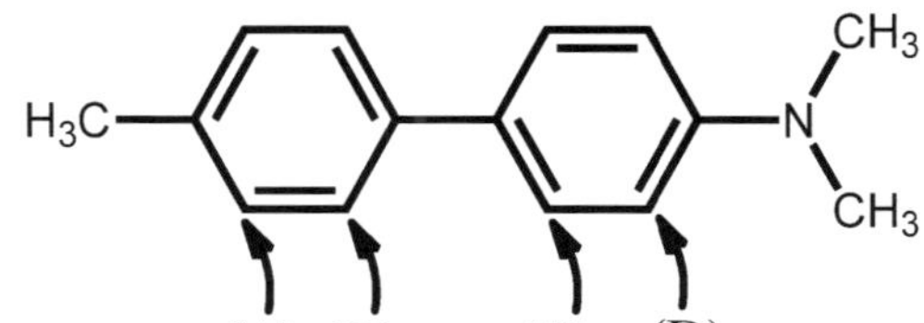

(A) (B) (C) (D)

17. Which is the best multistep synthetic route to give 1,3-diethylbenzene? (Starting with benzene)

(A) $\xrightarrow[AlCl_3]{CH_3CH_2Cl}$ $\xrightarrow[AlCl_3]{CH_3CH_2Cl}$ $\xrightarrow[NaOH,\ \Delta]{N_2H_4}$

(B) 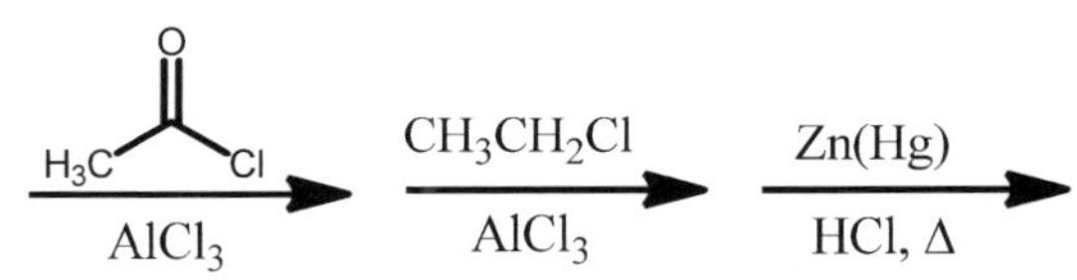

(C) $\xrightarrow[AlCl_3]{CH_3CH_2Cl}$ $\xrightarrow[AlCl_3]{CH_3CH_2Cl}$ $\xrightarrow[Pd]{H_2}$

(D) $\xrightarrow[H_2SO_4]{HNO_3}$ $\xrightarrow[AlCl_3]{CH_3CH_2Cl}$ $\xrightarrow[Pd]{H_2}$

18. Provide the product of this diazotization reaction:

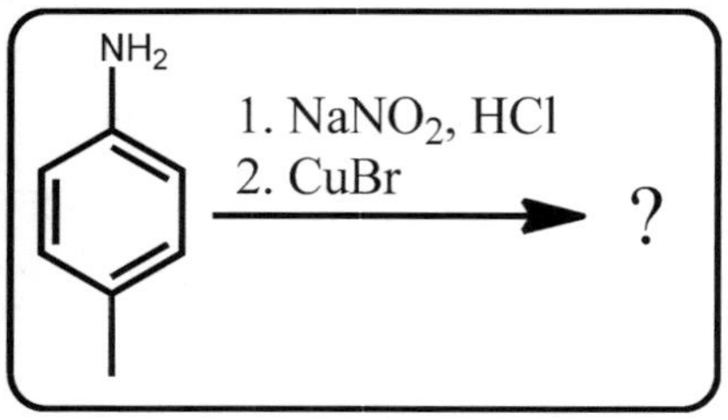

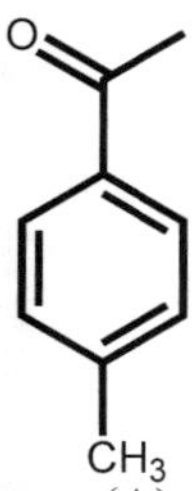

(A)

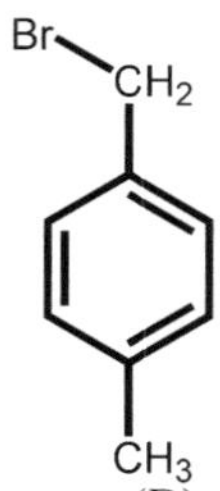

(B)

(C)

(D)

19. Which reaction will involve a **benzyne** intermediate?

(A)
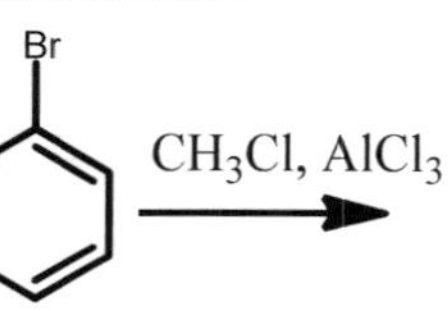

(B)
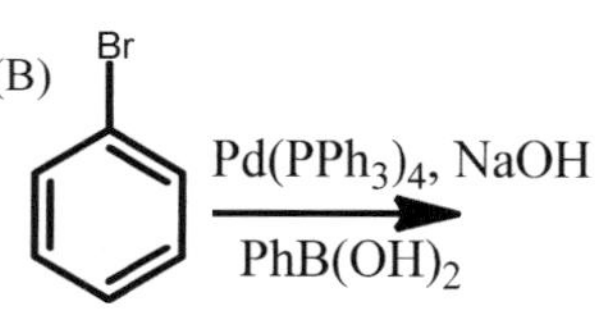

(C)
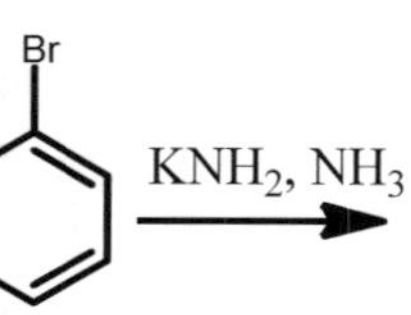

(D)
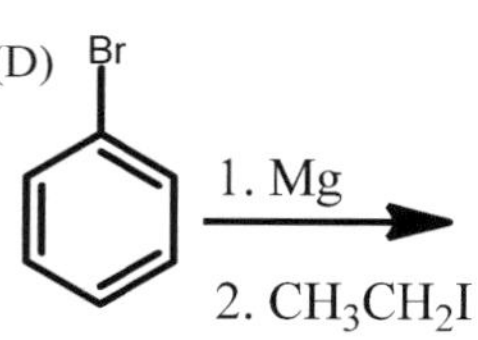

20. Which of these reactions will utilize carbocation rearrangement en route to the final major product?

(A)
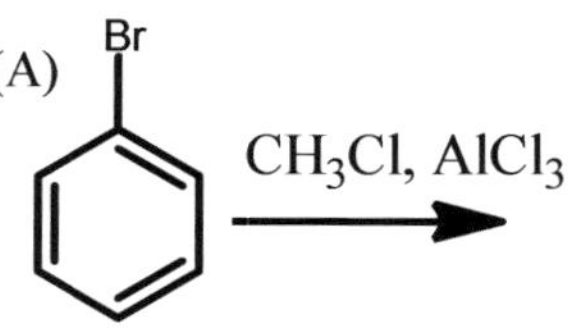

(B)
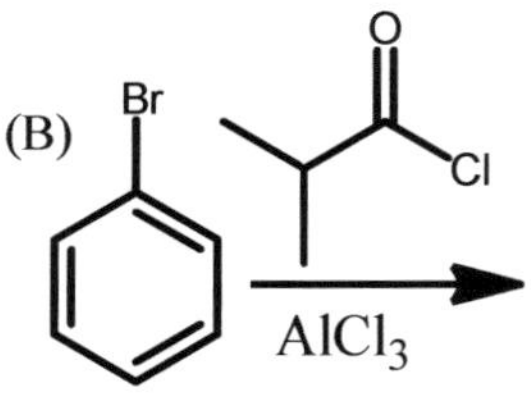

(C)
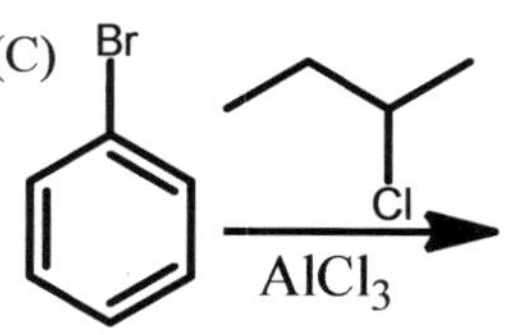

(D)
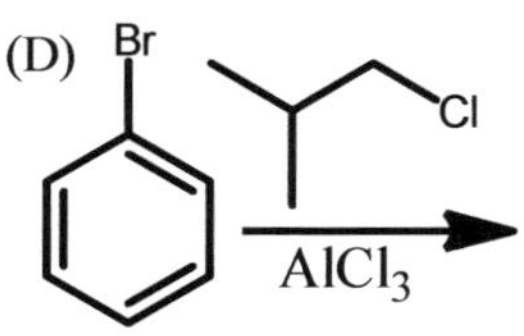

For numbers 21-22, consider the S_N2 reaction:

21. Which R-Br reacts fastest?

(A)

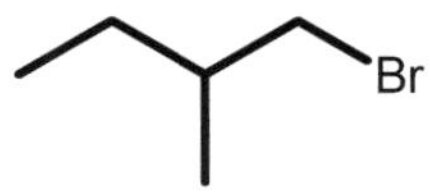

(B)

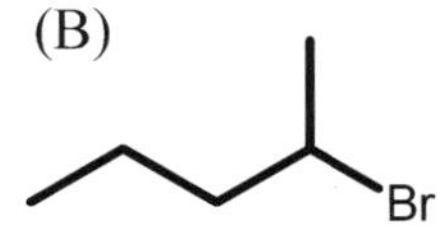

(C)

(D)

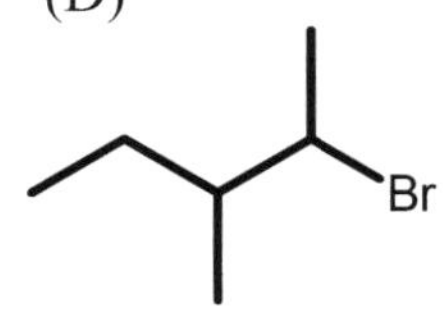

22. Which reaction coordinate diagram corresponds to the S_N2 reaction?

(A)

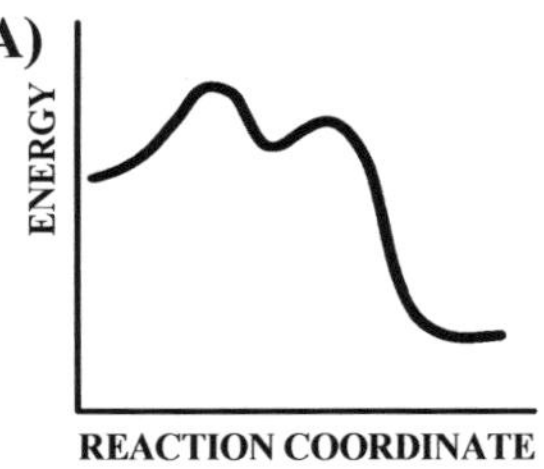

(B)

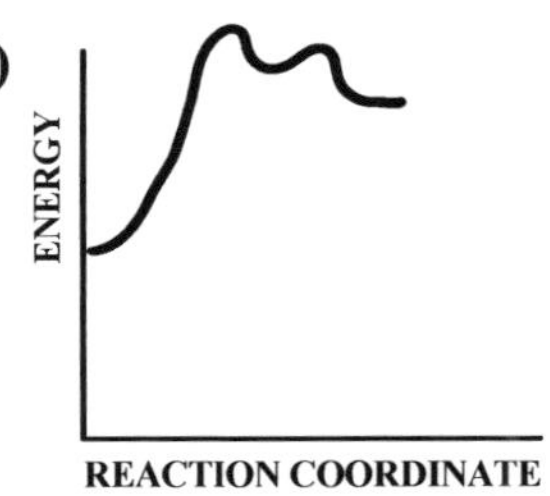

(C)

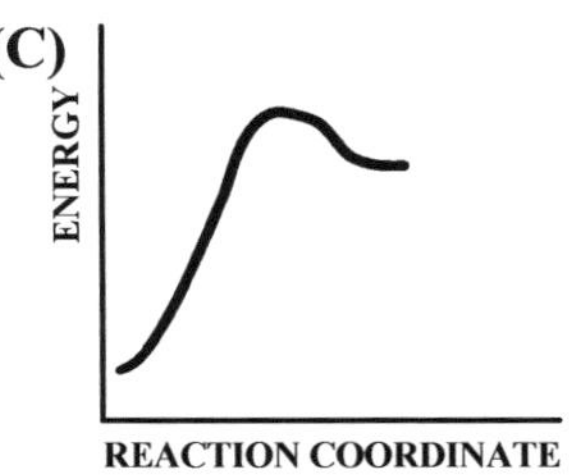

(D)

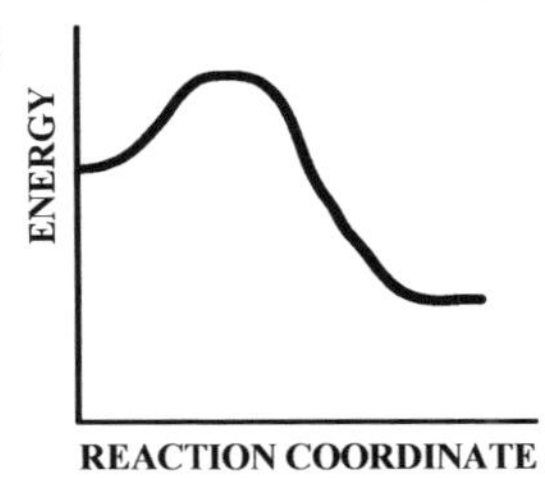

For numbers 23-24, consider the S_N1 reaction:

23. Which R-Br reacts fastest?

(A)

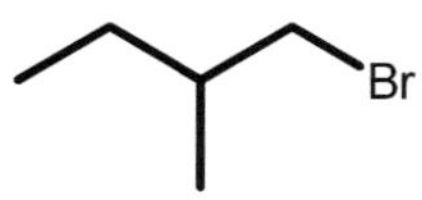

(B)

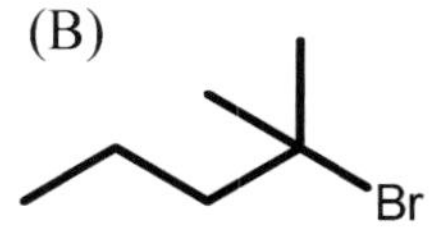

(C)

(D)

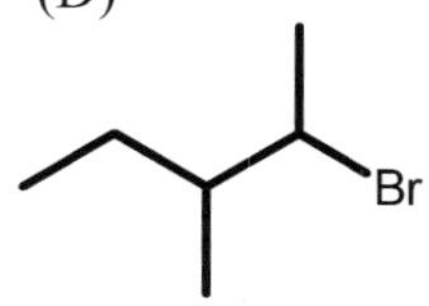

24. Which reaction coordinate diagram corresponds to the S_N1 reaction?

(A)

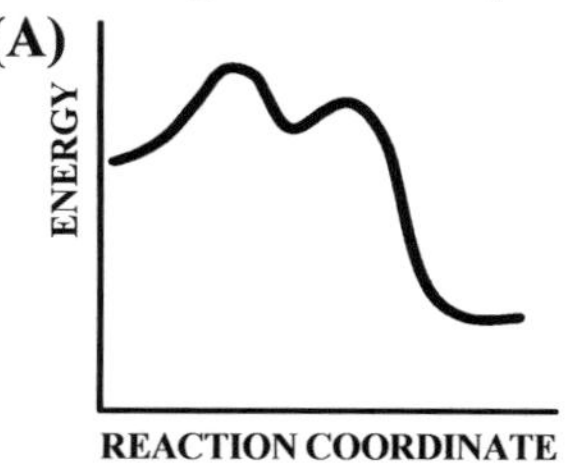

(B)

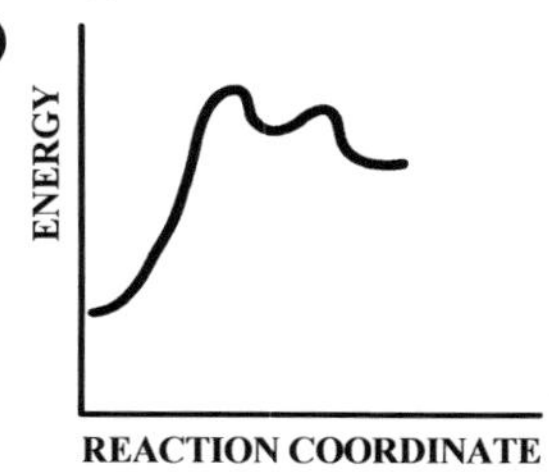

(C)

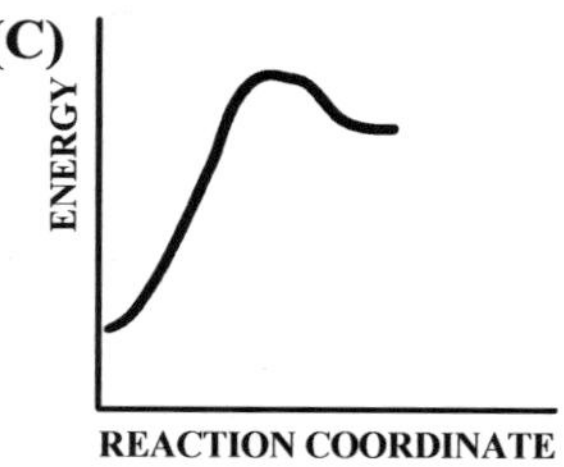

(D)

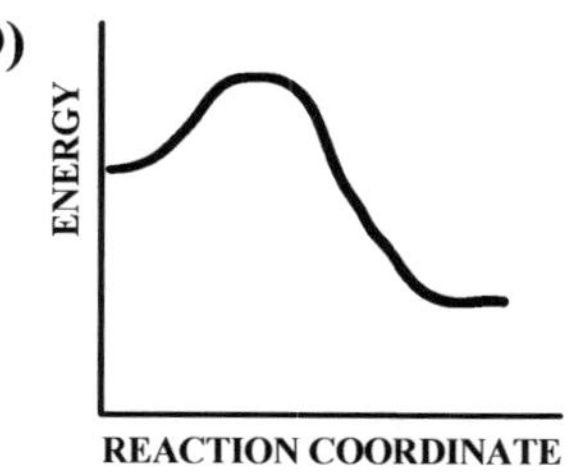

25. Provide the major product of the following E2 Reaction:

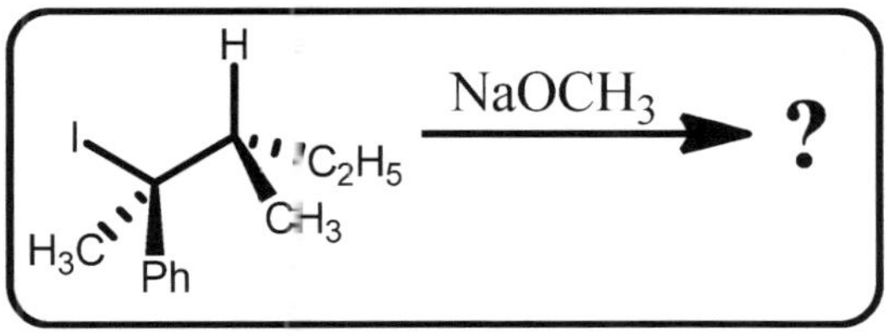

(A)

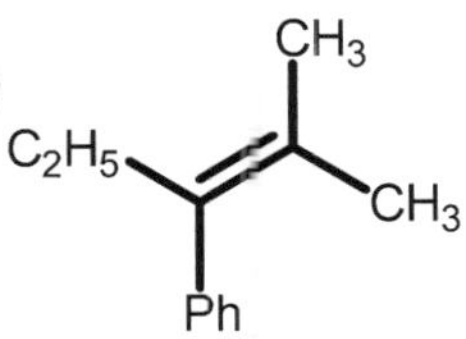

(B)

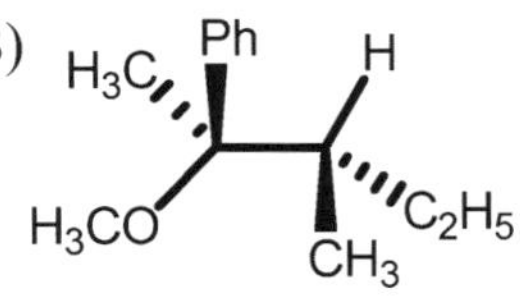

(C)

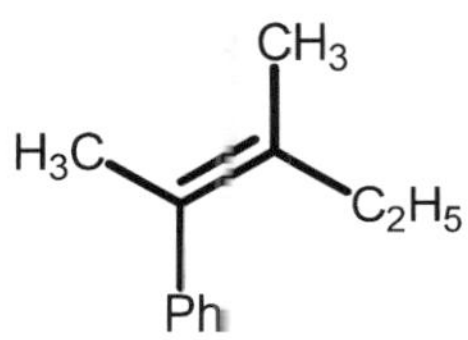

(D)

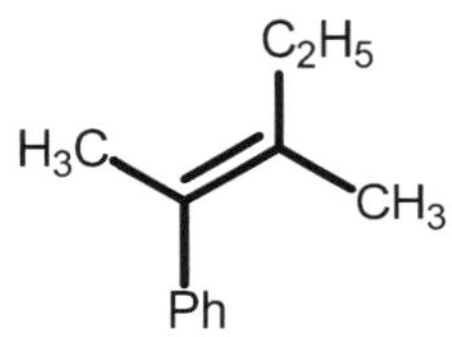

26. Which of the following reactions would give the best yield of 2-methylcyclopentanol?

(A)

$\xrightarrow[H_2O]{H_2SO_4}$

(B)

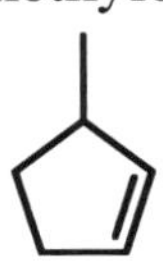

$\xrightarrow[H_2O]{H_2SO_4}$

(C)

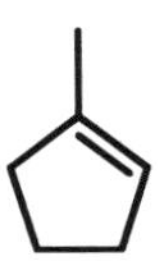

1. BH_3 → 2. H_2O, H_2O_2, NaOH

(D)

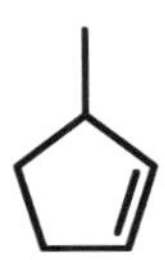

1. $Hg(OAc)_2$, H_2O → 2. $NaBH_4$

27. Consider the reaction:

ALKENE $\xrightarrow[2.\ NaHSO_3]{1.\ OsO_4}$ *Syn*-DIOL

What would be the major product if the alkene starting material is cyclopentene?

(A)

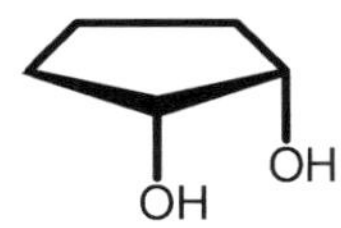

(B)

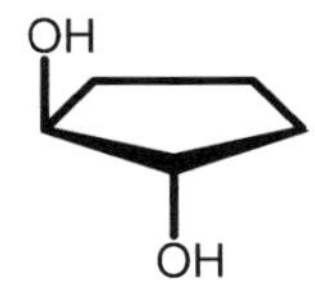

(C)

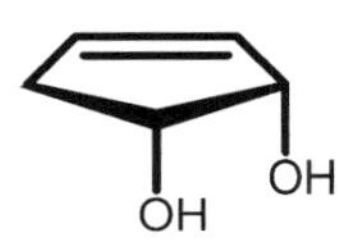

(D)

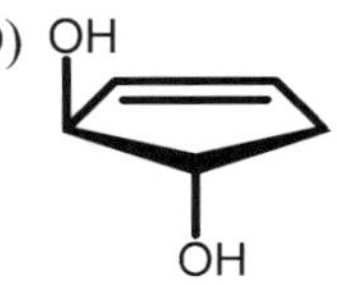

28. Which reaction sequence would give the best yield of the target product?

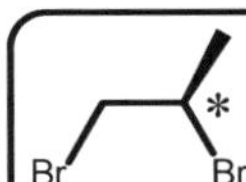

(specific stereoisomer, not a racemic mixture!)

(A)

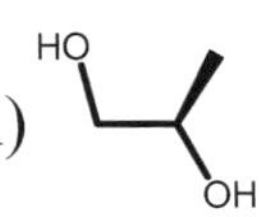

$\xrightarrow{2\ ClTos}$ $\xrightarrow{2\ NaBr}$

(B)

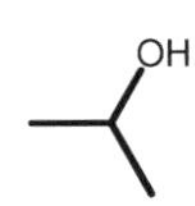

$\xrightarrow{H_2SO_4}$ $\xrightarrow{Br_2}$

(C)

$\xrightarrow{2\ HBr}$ $\xrightarrow{PBr_3}$

(D)

$\xrightarrow{2HBr}$ $\xrightarrow{PBr_3}$

29. Grignard reagents are strong bases. Of the following reagents, which one would **not** be deprotonated by a Grignard reagent?

(A)

(B)

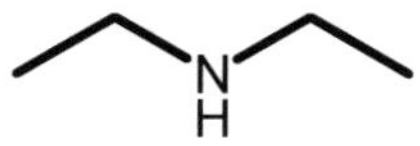

(C)

(D)

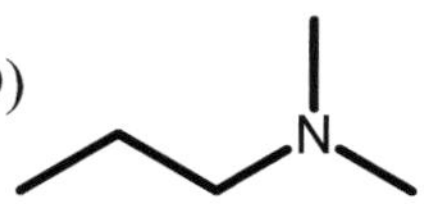

30. Grignard reagents can be very good nucleophiles. On this basis, which site would a Grignard reagent attack in the following molecule?

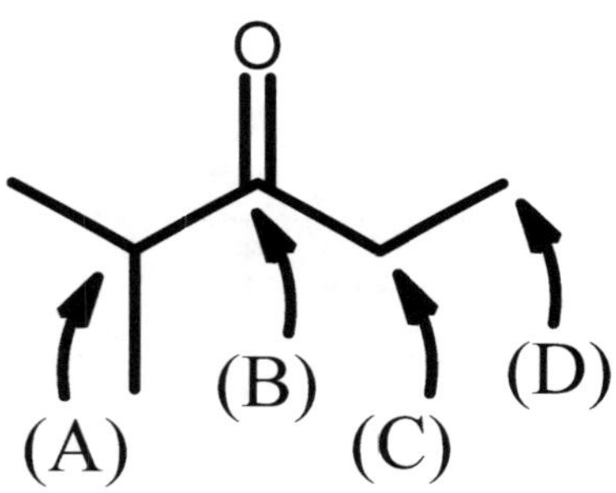

Key for Progress Check 1:

1. B
2. D
3. C
4. A
5. B
6. D
7. D
8. C
9. A
10. D
11. B
12. B
13. A
14. A
15. B
16. D
17. B
18. C
19. C
20. D
21. C
22. D
23. B
24. A
25. C
26. C
27. A
28. C
29. D
30. B

Detailed Feedback for Progress Check 1

For each question that you missed or were unsure of, review the Necessary Knowledge and Reasoning and possible mistakes provided below. By addressing the gaps in your knowledge and avoiding similar mistakes on future problems, you should see improvement in your scores.

1. *Necessary Knowledge and Reasoning:*

A quick way to assign hybridization to an atom is to figure out the sum of its substituents (attached atoms + lone pairs). Remember that lone pairs are not always drawn in a structure, so be sure to draw in any lone pairs needed to give the atom the formal charge shown in the structure that you are examining. Once you find the sum of its substituents (attached atoms + lone pairs), you assign hybridization as:

If (attached atoms + lone pairs) = 4, then hybridization = sp^3
If (attached atoms + lone pairs) = 3, then hybridization = sp^2
If (attached atoms + lone pairs) = 2, then hybridization = sp

The answer is B.

2. *Necessary Knowledge and Reasoning:*

a. What the *trans*- or *cis*- designates in a name.

b. The sites pointing straight up or straight down on cyclohexane are axial sites, whereas those that are drawn slanting up or down at angles are equatorial sites.

c. On cyclohexane, a group larger than hydrogen should be in an equatorial site to avoid steric strain.

d. If there is a choice between two larger-than-hydrogen substituents, the larger one is equatorial.

e. The molecule with the least strain is the most stable molecule.

The answer is D.

3. *Necessary Knowledge and Reasoning:*

a. The *weakest* acid is the one that produces the *least* stable anion upon removal of the proton.

b. A more electronegative element loses its proton more easily (in this case the H on the O is removed rather than an H on any of the C atoms.

c. If the negative charge can be delocalized via resonance, it acquires greater stability as compared to a molecule lacking resonance delocalization but has the negative charge on the same atom.

d. If there are more branches on the atom to which the anionic atom is attached, this will destabilize the anion be repulsive inductive effect.

The answer is C.

4. *Necessary Knowledge and Reasoning:*

a. Water is neither a good nucleophile nor a strong base.

b. An alkyl halide undergoes S_N1 and/or E1 reactions under such conditions.

c. An E1 reaction would yield an alkene (molecule with a C-C double bone) as a product. None of the choices are alkenes, so we only focus on S_N1 products.

d. The first step of an S_N1 reaction is formation of the carbocation upon leaving group removal.

e. If there is a more substituted site *directly* next to the positively-charged carbon of the carbocation, the carbocation will rearrange to place the positive charge on the adjacent, more substituted carbon.

f. Once the final most stable carbocation is identified, we need to know that the OH part from water will end up attached to the positively charged carbocation.

The answer is A.

5. *Necessary Knowledge and Reasoning:*

a. The Suzuki reaction is a C-C bond forming reaction.

b. The C-C bond that forms is between the carbon with the boron on it and the carbon with the leaving group on it (here the leaving group is Br).

c. There is no rearrangement during the reaction, so the boron and Br must be on the correct atoms to make the target molecule.

Here, we choose between various B-containing molecules. In the product, there is a methyl group *ortho-* to where the boron was, and another methyl group *meta-* to where the boron was.

The answer is B.

6. *Necessary Knowledge and Reasoning:*

a. The Diels-Alder reaction takes a diene and a dienophile (something with a C-C pi bond) and creates a six membered ring. If the dienophile is an alkene, the six-membered ring is a cyclohexene:

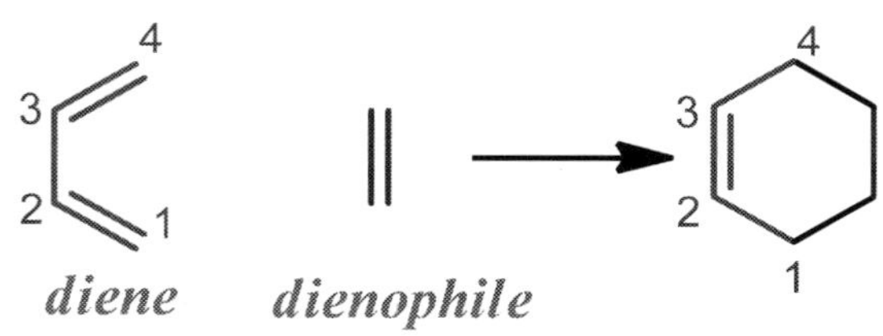

b. If there are substituents on the dienophiles, they will orient the same way with respect to one another in the product as they do in the starting material. For example, in the following reaction A and A' are *cis-* to one another in the starting dienophile as well as being *cis-* to one another in the product:

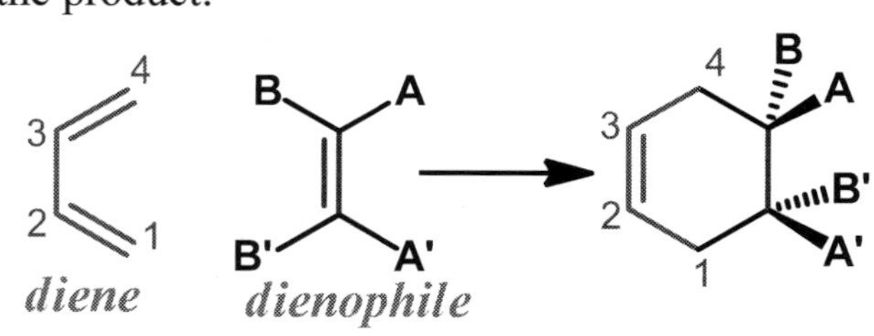

c. If there is a substituent on the diene, it will 'bend out of the way' upon reaction and attain the correct tetrahedral geometry about the sp^3 carbons (numbers 1 and 4 in the scheme below):

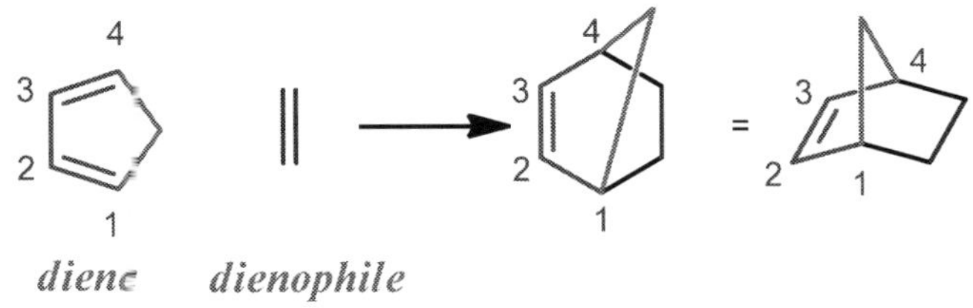

The answer is D.

7. *Necessary Knowledge and Reasoning:*

a. Water is neither a good nucleophile nor a strong base.

b. An alkyl halide undergoes S_N1 and/or E1 reactions under such conditions.

c. An E1 reaction would yield an alkene (molecule with a C-C double bone) as a product. None of the choices are alkenes, so we only focus on S_N1 products.

d. The first step of an S_N1 reaction is formation of the carbocation upon leaving group removal.

e. A primary carbocation is too unstable to form in this S_N1 reaction.

f. Because the intermediate cannot form, the reaction cannot proceed.

The answer is D.

8. *Necessary Knowledge and Reasoning:*

a. Water is a poor nucleophile (it is neutral and is not ionic, so it will not fully dissociate to ions). It is also a weak base.

b. When a poor nucleophile that is a weak base reacts with a secondary alkyl halide, the mechanism is expected to be formation of a carbocation, which can then do an S_N1 or E1 reaction.

c. In this case the question only asks for the substitution product (NOT elimination), so we only have to worry about the S_N1 pathway.

d. In an S_N1 mechanism the product has a mixture of configurations is substitution happens at the stereogenic carbon. Here, this leads to an approximately 50:50 ratio of both *R*- and *S*-isomers, which is called a racemic mixture.

The answer is C.

9. *Necessary Knowledge and Reasoning:*

a. The starting material is an epoxide. Epoxides are attacked by nucleophiles at the less substituted carbon under basic conditions. Epoxides are attacked on the more substituted carbon under acidic conditions.

b. The first condition that the epoxide is exposed to includes H_2SO_4 (sulfuric acid), so the nucleophile will attack the more substituted side of the molecule.

c. The nucleophile is $HOCH_3$, so a $–OCH_3$ unit will end up attached to the more substituted carbon of the epoxide ring.

d. The initial reaction product will thus be:

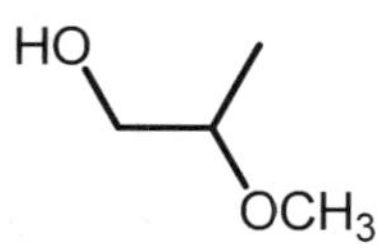

Note that although the carbon on the right is a stereogenic (chiral) center, the stereochemistry of the starting material is not specified, so we cannot determine a specific configuration in the product either.

e. The next step is to react this initial product with PBr_3.

f. PBr_3 reacts with alcohols to replace the OH with a Br.

g. The mechanistic step wherein the Br attaches is an S_N2 step.

h. There is no intermediate in an S_N2 reaction and thus no rearrangement.

i. The S_N2 reaction works well on primary sites.

j. The final product is thus:

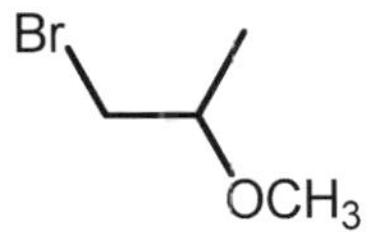

The answer is A.

10. *Necessary Knowledge and Reasoning:*

a. In addition reactions, alkenes tend to donate the pi bond electrons to the electron deficient portion of the reactant in order to create an intermediate.

b. In solution, HBr (the most reactive reactant in choice A or B) dissociates to form H^+ and Br^-.

c. When an alkene donates electrons to the H^+, a **carbocation intermediate** results:

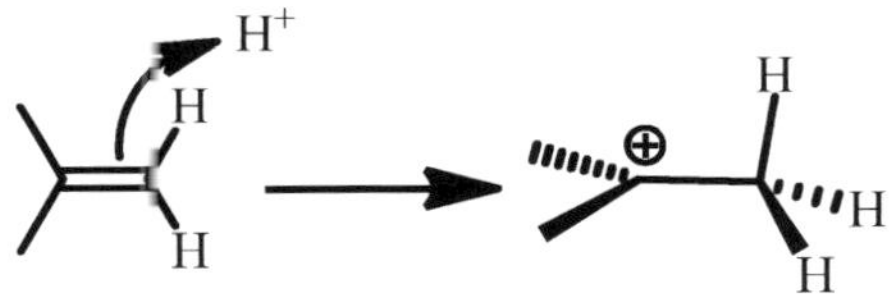

This rules out A and B as involving a bromonium ion.

d. In choice C, the alkene will attack the I end of the polar reactant I-Cl in step A to form the iodonium ion intermediate. Then the chloride attacks in an S_N2 reaction (step B):

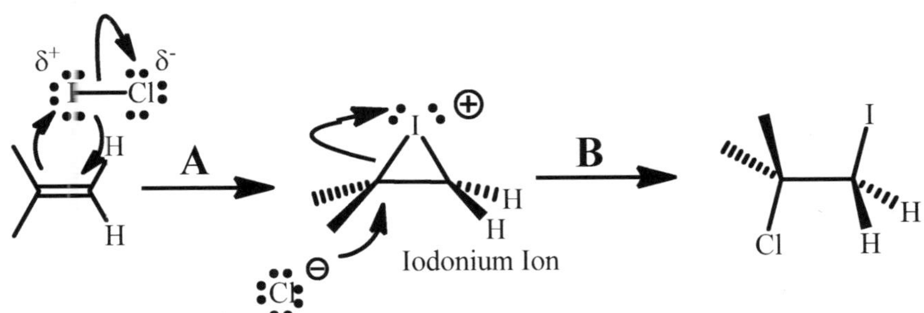

e. When NaBr is in solution it forms Na^+ and Br^-. The bromide is a good nucleophile and simply displaces the best leaving group (I) in an S_N2 reaction (concerted, so no intermediate). So, though the reaction in part C is a bit involved, it has no bromonium ion intermediate. Therefore, we can rule it out.

f. Alkenes react with bromine (Br_2 in choice D) in the bromination reaction to form the bromonium intermediate:

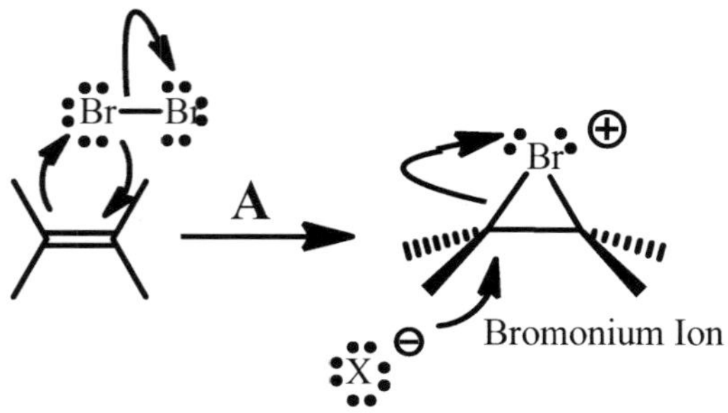

The answer is D.

11. *Necessary Knowledge and Reasoning:*

a. To be aromatic, a molecule must be cyclic. All choices are cyclic.

b. If a cyclic molecule is aromatic, it will attain coplanarity of all the atoms in the pi system.

c. There must be no sp^3-hybridized atoms in the cycle if it is to be aromatic.

d. There must be a total of $4n+2$ electrons (where n is an integer) in the pi system for the molecule to be aromatic (i.e., number of electrons = 2, 6, 10, 14, 18, 22 … etc.).

e. Only choice B has the right number of pi system electrons: two from each of the two pi bonds and two from the lone pair = six total. Solving for n in $4n+2 = 6$ gives $n = 1$, an integer.

The answer is B.

12. *Necessary Knowledge and Reasoning:*

a. To be aromatic, a molecule must be cyclic. All choices are cyclic.

b. If a cyclic molecule is aromatic, it will attain coplanarity of all the atoms in the pi system.

c. There must be no sp^3-hybridized atoms in the cycle if it is to be aromatic.

d. There must be a total of $4n+2$ electrons (where n is an integer) in the pi system for the molecule to be aromatic (i.e., number of electrons = 2, 6, 10, 14, 18, 22 … etc.).

e. Choice A has the right number of pi system electrons: two from the pi bond. Solving for n in $4n+2 = 2$ gives $n = 0$, an integer.

f. Choice C also has the right number of pi system electrons: two from each of the two pi bonds and two from the lone pair = six total. Solving for n in $4n+2 = 6$ gives $n = 1$, an integer.

g. Because the question asks for the one that is **not** aromatic and choice B has the **wrong** number of electrons, this is the correct answer.

The answer is B.

13. *Necessary Knowledge and Reasoning:*

a. A Lewis Base is a lone pair donor.

b. In order for a nitrogen to be effective at donating its lone pair, the lone pair cannot be tied up in resonance or in an aromatic system.

c. If an sp^3 hybridized atom with a lone pair is adjacent to an atom that is engaged in a pi bond (a double or triple bond), it will be tied up in a resonance structure and thus the nitrogen will not be basic.

d. If an atom with a lone pair is sp^2 hybridized, it cannot donate its lone pair to an adjacent pi system, so the lone pair is available for donation to an acid.

e. A nucleus has a greater attraction for electrons in an sp^2 orbital than in an sp^3 orbital, so it is easier for an sp^3 hybridized atom to donate its electrons (which is necessary for it to act as a base).

The answer is A.

14. *Necessary Knowledge and Reasoning:*

a. Electrophilic aromatic substitution reactions utilize a carbocation intermediate.

b. Resonance donors are the best substituents for stabilizing the carbocation intermediate and thus speed up the reaction.

c. The best resonance donors are nonhalogen substituents wherein a lone pair is on the atom directly attached to the benzene ring.

The answer is A.

15. *Necessary Knowledge and Reasoning:*

a. The structure of the tosyl group is:

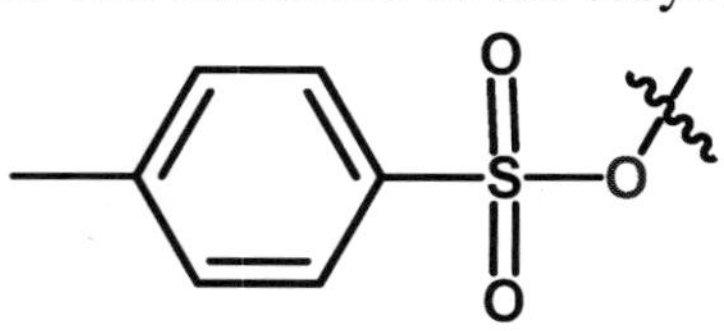

b. If this group left the molecule to which it is attached, it becomes the tosylate anion, the structure of which is:

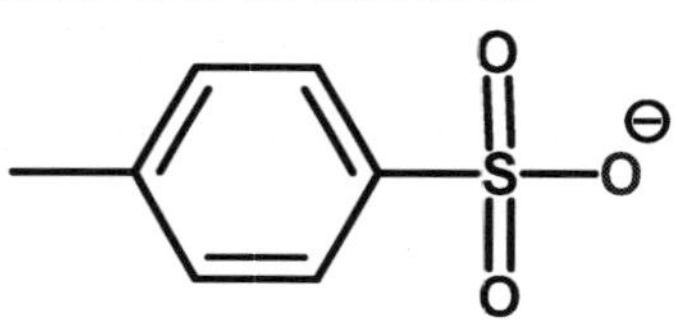

c. The tosylate anion is very stable because the anion is on an electronegative element (O) and there are three resonance structures that can be drawn wherein the negative charge is spread out over three oxygen atoms.

d. A strong base (choice A) would be an unstable anion, so neither the tosyl group (as the question asked) nor the tosylate anion is a strong base. This rules out choice A.

e. In order to evaluate whether the tosyl group is good nucleophile (choice C), we need to recall the trends in nucleophilicity. Look at the atom on which the negative charge is positioned. The nucleophilicity increases as the size of the atom increases. If the atoms one is comparing are the same size (in the same row) the nucleophilicity trend follows the same trend as basicity.

f. If an anion is bulky it is not a good nucleophile. As a general rule, a nucleophile is considered bulky if the anionic-atom-adjacent site has three or more atoms on it, such as:

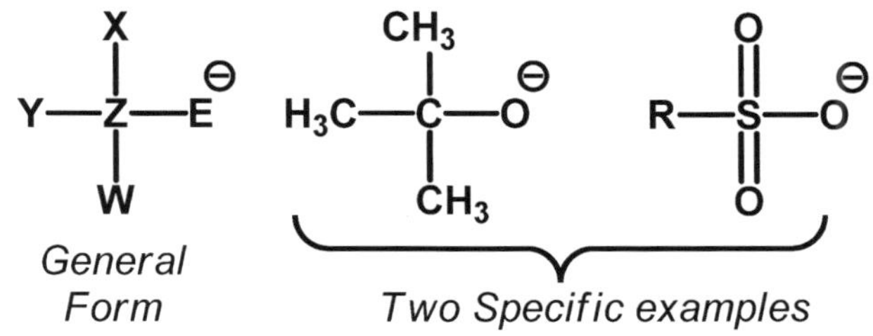

g. The tosyl group is not anionic. Even the tosylate anionic form of the tosyl group is not a good nucleophile because it is bulky. This rules out choice C.

h. Choice D asks whether the tosyl group is an acid. The tosyl group does not have an acidic proton, so it cannot act as a Bronsted-Lowry or Arrhenius acid. The tosyl group also lacks an empty orbital to accept a lone pair, so it cannot be a Lewis acid. This rules out choice D.

i. A good leaving group is something that forms a stable anion when it leaves the atom to which it is attached. A good general rule is that if an anion is more stable than fluoride or if the anion has resonance forms, we can consider it a good leaving group. The tosyl group forms the very stable tosylate anion when it leaves, so the tosyl group is a good leaving group.

The answer is B.

16. *Necessary Knowledge and Reasoning:*

a. Electrophilic aromatic substitution reactions utilize a carbocation intermediate.

b. Resonance donors are the best substituents for stabilizing the carbocation intermediate and thus speed up the reaction.

c. The best resonance donors are nonhalogen substituents wherein a lone pair is on the atom directly attached to the benzene ring.

d. The ring on the right, which has sites C and D on it, has a dimethylamino group (a resonance donor) on it, so this ring reacts faster than the other ring.

e. Resonance donors stabilize the cationic intermediate to a greater extent when the electrophile adds *ortho-* to the resonance donor than when the electrophile adds to the site *meta-* to the resonance donor.

The answer is D.

17. *Necessary Knowledge and Reasoning:*

a. **IF** a benzene has a substituent on it that is

i. Activating (nonhalogen group featuring a lone pair on the element directly attached to the benzene ring),

ii. Slightly Activating (a hydrocarbon), or

iii. Slightly Deactivating (a halogen),

THEN the next group to be added via an electrophilic aromatic substitution reaction will add *ortho-* or *para-* to the first substituent (a mixture of major products is observed).

b. One of conditions among the answer choices are an alkyl chloride (RCl) and aluminum trichloride ($AlCl_3$). These conditions are those for the Friedel-Crafts alkylation, wherein an alkyl group will add to the benzene ring.

c. In both choices A and C, the first reaction adds an ethyl group. The second reaction is another Friedel-Crafts reaction. However, the second group will add *ortho-* and *para-* to the first ethyl group, whereas the target product we are asked to make has the two ethyl groups *meta-* to one another. This rules out choices A and C.

d. As we move on to choice B, the first conditions are $AlCl_3$ and RC(O)Cl. These conditions are those of the Friedel-Crafts acylation.

e. The Friedel-Crafts acylation reaction adds an acyl group to an aromatic starting material.

f. In an acyl group, there is a partial positive charge on the carbon that is adjacent to the benzene ring:

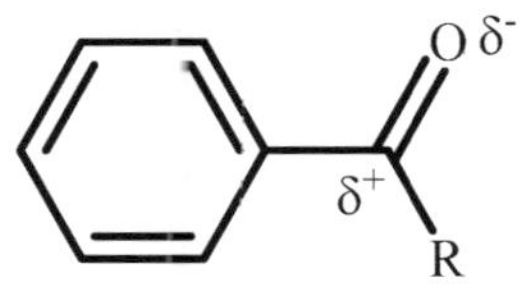

g. If there is a partial positive charge adjacent to the benzene ring, this deactivates the benzene ring to reactions that would put a positive charge on the benzene (like electrophilic aromatic substitution that has a carbocation intermediate). The acyl group is therefore an example of a Deactivating group for electrophilic aromatic substitution.

h. Consider the carbocation intermediate formed during the electrophilic aromatic substitution of a benzene that already has an acyl group on it. The carbocation intermediate is *least* stable when the electrophile adds *ortho-* or *para-* to the acyl group. For this reason, the *meta-*substitution is favored.

i. In choice B, a Friedel-Crafts Acylation first places an acetyl group on the benzene:

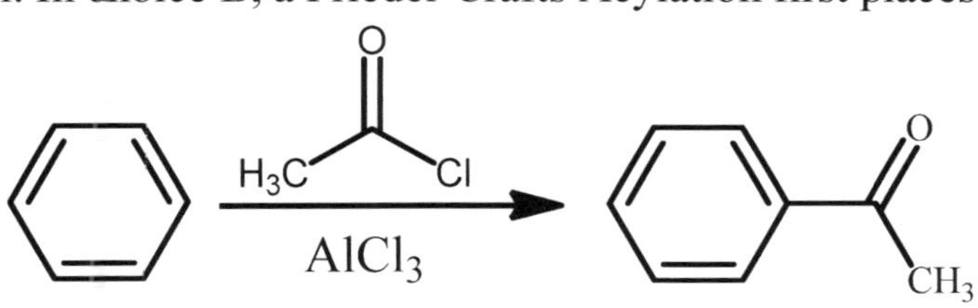

The next step is a Friedel-Crafts alkylation. The *meta*-directing ability of the acyl group causes the major product to be:

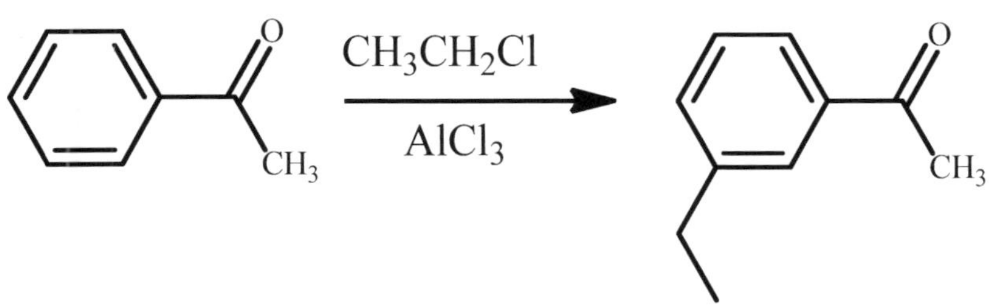

The third step in this choice is a Clemmensen reduction, which reduces a ketone to an alkane:

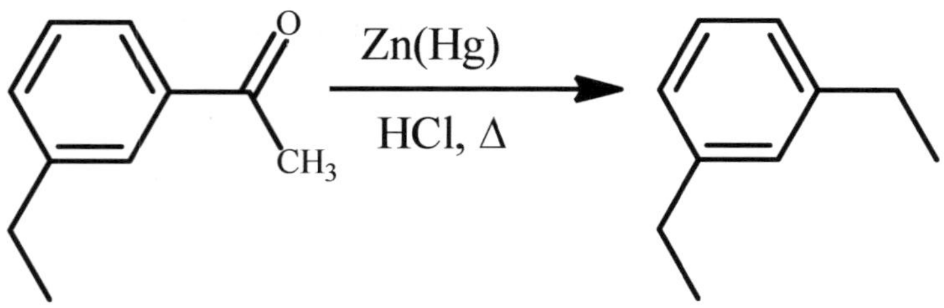

This route successfully yields 1,3-diethylbenzene.

j. Choice D places a nitro group (NO_2) on the ring, then an ethyl group *meta-* to it (nitro is a deactivating and *meta*-directing substituent like an acyl group), then the nitro group is reduced to an amino group (NH_2) by hydrogen and palladium. This does not lead to the right groups on the ring for the target product, so choice D is ruled out.

The answer is B.

18. *Necessary Knowledge and Reasoning:*

a. These first conditions form a diazo salt in place of the NH_2 group.

b. The diazo unit is an excellent leaving group (it leaves as N_2 gas), so it is readily replaced by the anionic portion of various metal salts.

c. Adding CuBr to a diazo salt places a Br in place of the diazo salt. The net result is that the NH_2 is substituted for a Br group.

The answer is C.

19. *Necessary Knowledge and Reasoning:*

a. A benzyne intermediate is usually drawn as a benzene ring with a triple bond in place of one of the double bonds:

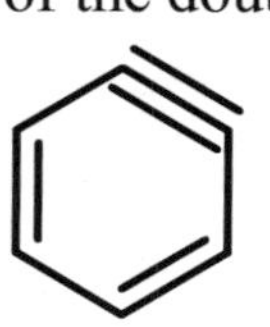

b. This intermediate results when a strong base removes a proton *ortho-* to a good leaving group.

c. Choice A is an electrophilic aromatic substitution reaction that involves a carbocation intermediate. This rules out choice A.

d. Choice B is a Pd catalyzed coupling reaction that involves organopalladium species as intermediates. This rules out choice B.

e. Choice C features a strong base (KNH_2 makes $^{-}NH_2$ in solution), which will remove the proton *ortho-* to Br (a good leaving group). This fits the criteria necessary for the formation of a benzyne intermediate and is the correct answer.

f. If we check choice D just to be sure, we see that the reaction does not use any strong bases. The first step (reaction with magnesium to form a Grignard reagent PhMgBr) is a simple oxidative addition. In the second step, the Grignard reagent will deploy Ph^{-} as a nucleophile in an S_N2 reaction (concerted, no intermediate).

The answer is C.

20. *Necessary Knowledge and Reasoning:*

a. Reaction choice B is a Friedel-Crafts acylation, which involves an acylium ion intermediate. The acylium ion does not rearrange, so this rules out choice B.

b. Reaction choices A, C and D involve reacting benzene with RCl and $AlCl_3$. These are the conditions of Friedel-Crafts alkylation.

c. The Friedel-Crafts alkylation reaction involves a carbocation intermediate.

d. The carbocation that initially forms is one on which the positive charge is on the carbon that had the Cl on it.

e. If there is a more substituted site *directly* next to the positively-charged carbon of the carbocation, the carbocation will rearrange to place the positive charge on the adjacent, more substituted carbon.

f. Choice A will form a methyl carbocation. There is only one carbon, so the carbocation will not be able to rearrange. This rules out choice A.

g. Choice C has the Cl on a secondary carbon, so the carbocation that initially forms will be a secondary carbocation:

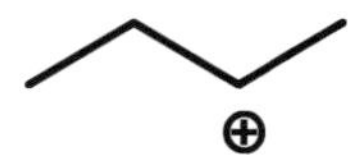

The carbon to the right of the positively charged center is a primary carbon, so no stability is added by shifting the charge there.

The carbon to the left of the positively charged center is a secondary carbon, so no stability is added by shifting there. Even if a shift were to occur to this position, the carbocation formed would be structurally identical, leading to no net change.

This rules out choice C.

h. Choice D has the Cl on a primary carbon, so the carbocation that initially forms will be a primary carbocation:

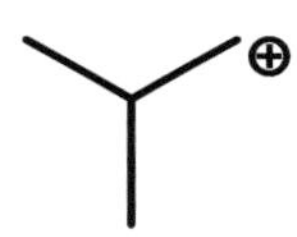

This carbocation is next to a tertiary carbon, so a 1,2-hydride shift to move the carbocation to that site will lead to a much more stable tertiary carbocation. This means that the shift will happen in this case.

The answer is D.

21. *Necessary Knowledge and Reasoning:*

a. The S_N2 reaction involved nucleophilic attack at the carbon that has a leaving group on it at the same time that the leaving group leaves as an anion. The process is concerted (one single step), so there are no intermediates.

b. The nucleophile will more quickly attack a carbon that is less sterically hindered.

c. A primary site is less sterically hindered than a secondary site, which is less sterically hindered than a tertiary site.

d. Choices A and C each have the leaving group (Br) on a primary carbon. This means that they will react faster than the substrates in choices B and D.

e. Choice A, however, has some steric bulk (a branch) on the carbon *adjacent to* the carbon with the leaving group on it:

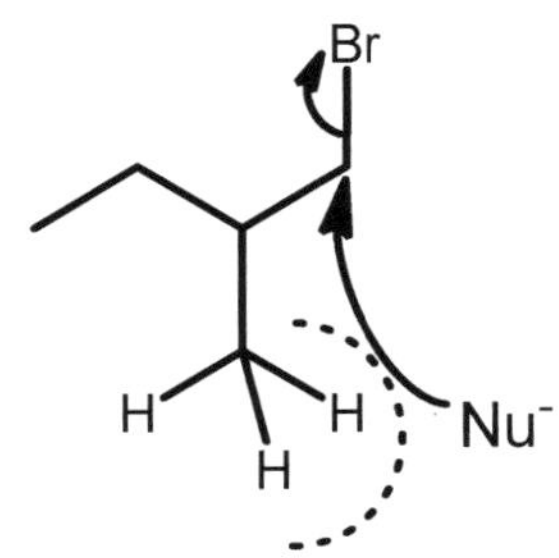

A branch on a carbon *adjacent* to the carbon with the leaving group on it is called a β-branch. This bulk slows down the nucleophile's attack compared to choice C, which does not have a β-branch. This rules out choice A.

The answer is C.

22. *Necessary Knowledge and Reasoning:*

a. The S_N2 reaction is concerted.

b. The reaction coordinate diagram for a concerted reaction should show a transition state but not an intermediate (there are no intermediates in a concerted reaction).

c. On a reaction coordinate diagram, a transition state is observed as a peak, whereas an intermediate is observed as a valley.

d. If the reactants are less stable than the products, then the equilibrium will lie more towards the products' side. Therefore, the reactants will need to be less stable (higher in energy) for this reaction to proceed in the desired direction.

e. Only the reaction coordinate diagram in choice D fits the criteria for the S_N2 reaction.

The answer is D.

23. *Necessary Knowledge and Reasoning:*

a. The S_N1 reaction mechanism involves a carbocation intermediate.

b. A carbocation is stabilizeds via *hyperconjugation* with any adjacent hydrocarbon substituents, so the stability of carbocations follows the general trend:

tertiary > secondary > primary > methyl

c. The more stable the carbocation intermediate, the faster the reaction.

The answer is B.

24. *Necessary Knowledge and Reasoning:*

a. The S_N1 reaction features a carbocation intermediate.

b. The reaction coordinate diagram must show the formation of the intermediate prior to the formation of product.

c. On a reaction coordinate diagram, a transition state is observed as a peak, whereas an intermediate is observed as a valley.

d. If the reactants are less stable than the products, then the equilibrium will lie more towards the products' side. Therefore, the reactants will need to be less stable (higher in energy) for this reaction to proceed in the desired direction.

e. Only the reaction coordinate diagram in choice A fits the criteria for the S_N1 reaction.

The answer is A.

25. *Necessary Knowledge and Reasoning:*

a. The E2 reaction involves concerted removal of a leaving group from one carbon and a proton from the adjacent carbon. This leads to formation of a pi bond between the two carbons from which the two groups have been removed.

b. The H that is removed will be the one on the most substituted carbon (remember that the only carbon choices are carbons *adjacent* to the carbon with the leaving group on it).

c. The H and leaving group must be antiperiplanar, meaning that they are *anti-* to one another in a Newman projection looking down the C-C bond that becomes the double bond.

d. In the starting material provided, the indicated H must be removed:

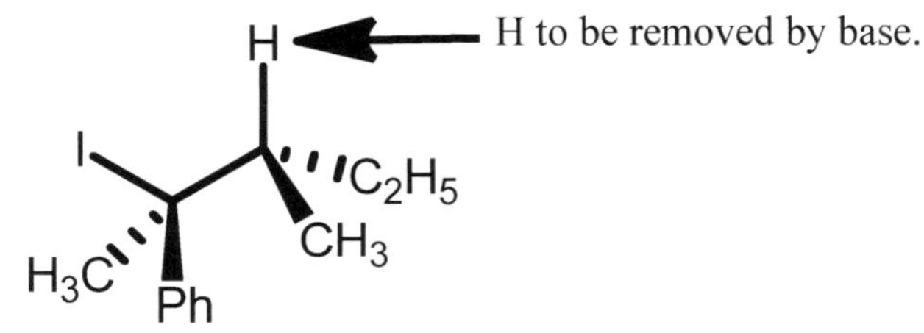

e. As drawn, the H and I (leaving group) are not antiperiplanar. We have to rotate 180° about the C-C bond indicated in order for the reaction to take place:

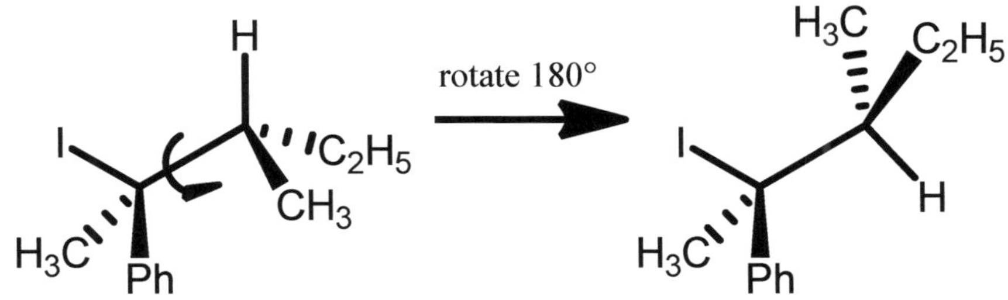

f. Because the E2 reaction is concerted (no intermediate or rearrangement during reaction), any groups pointing in the same direction in the starting material (rotated to be ready to do reaction) will point in the same way in the product. This means that in the alkene product, the Ph and C_2H_5 groups (both pointed out of the page in the starting material on the right above) must point the same way. Also, the two methyl groups (both pointed into the page in the starting material on the right above) must point the same way in the product. The only alkene that meets these criteria is the one shown in choice C.

The answer is C.

26. *Necessary Knowledge and Reasoning:*

a. The structure of the target product, 2-methylcyclopentanol, is:

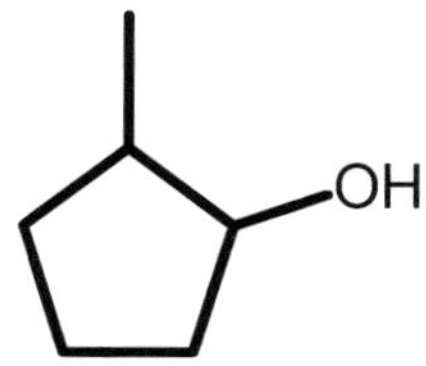

b. The conditions in choices A and B are conditions of the hydration reaction, which form a Markovnikov alcohol from an alkene (an OH and an H are added).

c. A Markovnikov product is one in which the more electronegative element is bound to the more substituted carbon (here this means that the OH will end up on the more substituted carbon).

d. The hydration reaction involves a carbocation intermediate that can rearrange prior to addition of the OH unit to the cationic center.

e. In choice A, one of the two carbons in the double bond will have an H added to it. After carbocation rearrangement, one of the other carbons will have an OH added to it. This does not specifically put the OH on the 2-carbon, so A is ruled out. A small amount of the 2-methylcyclohexanol may form prior to rearrangement, but it will not be a major product.

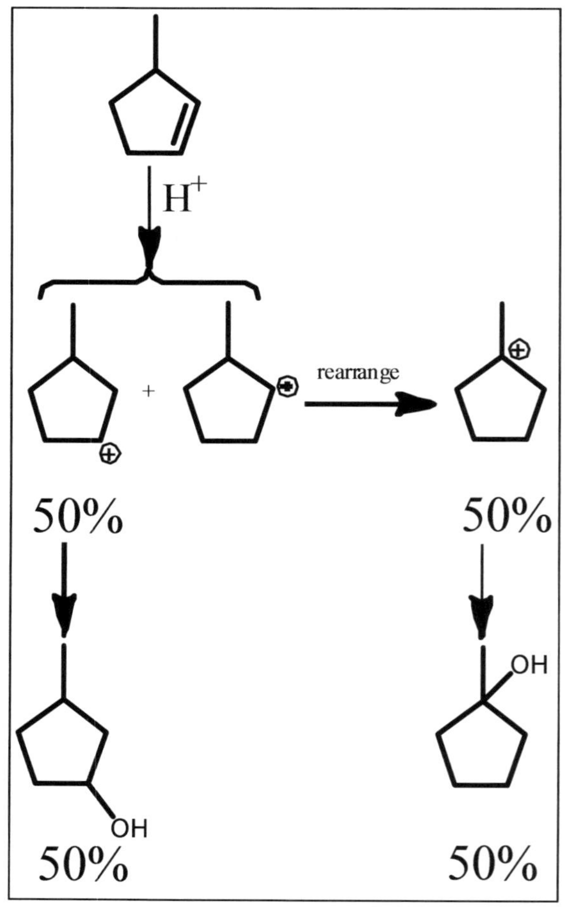

f. In B, both carbons in the double bond are secondary. So, about 50% of each carbocation will form. One of the carbocations will rearrange. Then the two carbocations go on to add the OH unit to the positively charged carbon as shown in the scheme to the right.

Neither of the major products, however, is the desired product. This rules out choice B.

g. The reaction in choice C is the hydroboration/oxidation. This reaction proceeds without rearrangement to make the anti-Markovnikov alcohol. This means that the major product will feature the OH on the less substituted of the carbons that are in the double bond of the starting material. This would be the product we are trying to make.

h. Checking choice D, we see that it is an oxymercuration / reduction. This reaction produces a Markovnikov alcohol, just like hydration does, but without the complication of carbocation intermediate rearrangement. So, we will get a product mixture with 50% of 2-methylcyclopentanol and 3-methylcyclopentanol in similar amounts. We are asked for the *best* yield of the target. Because this choice leads to a *mixture of two* major products whereas choice C leads to *one* major product. This makes D a poorer choice than choice C.

The answer is C.

27. *Necessary Knowledge and Reasoning:*

a. The question already tells you that this reaction makes a *syn*-diol with loss of the double bond. If we know what a *syn*-diol is, we can get the answer. The designator of *syn-* indicates that the two groups are on the same side as one another. Only choice A is a *syn*-diol without the double bond remaining.

The answer is A.

28. *Necessary Knowledge and Reasoning:*

a. In order to get a specific chiral isomer as product rather than a mixture of isomers, one of the starting materials or reagents must be a chiral molecule. This general premise lets us rule out choice B immediately.

b. If we examine choice A, the first step is a reaction that converts the OH groups to a good leaving groups, tosyl groups.

The second reaction will be an S_N2 reaction to place a Br on each of the carbons that has a tosyl group.

The initial starting material is in the right stereochemical arrangement. The S_N2 reaction, however, leads to Walden inversion of the stereochemistry, so the configuration of the product will be opposite of the desired stereochemistry. This lets us rule out choice A.

c. In choice C, we need to know that under acidic conditions the nucleophile will attack the more substituted side of the ring. This will yield as a product of the first step:

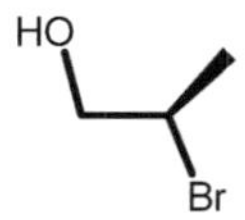

Note that the attack of the ring by bromide is an S_N2 reaction, so the configuration of the chiral carbon (the carbon on the right) is inverted so it is now in the correct configuration.

The second reaction uses PBr_3 to replace the OH on the left carbon with a Br. This carbon is not a chiral center so there is no concern about chirality changes. This will give the target product.

The answer is C.

29. *Necessary Knowledge and Reasoning:*

a. A proton will be predominantly removed by a base if and only if the anion produced is a more stable anion than the initial base anion:

This reaction is favorable in the forward direction.

b. The Grignard reagent can be thought of as producing a carbanion in solution.

c. A negative charge on a carbon atom is less stable than a negative charge on a more electronegative element like O or N, so a Grignard will deprotonate and H-O or H-N unit.

d. This question asks which will *not* be deprotonated by a Grignard reagent, so we are looking for the choice that does not have any H-O or H-N units. Only choice D meets this criterion.

The answer is D.

30. *Necessary Knowledge and Reasoning:*

a. A nucleophile is a species that is attracted to positive or partially positive centers.

b. A partial positive charge is formed on an atom when a more electronegative atom is attached to it.

c. In this example, there is only one carbon with an electronegative atom on it (the C with the double bond to O, choice B).

The answer is B.

Progress Check 2:

Material Coverage:

"Organic Chemistry 2 Lecture Guide 2013":

Lecture Sets 1 and 4-7

Date by which material should be studied:

Chapter(s) in text covering material:

Date this problem set is due:

Points this problem set is worth:

PROGRESS CHECK 2

1. Which is the ***strongest*** acid?

(A)

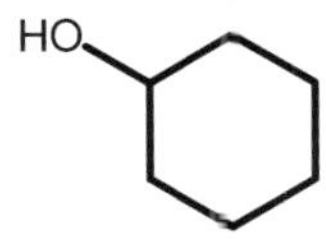

(B)

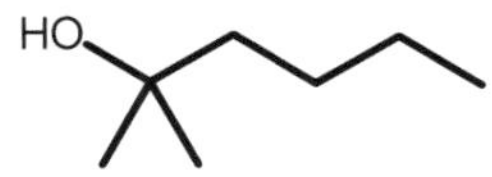

(C)

(D)

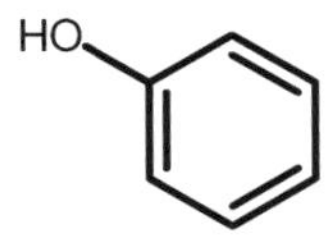

2. Which is the best nucleophile?

(A)

(B)

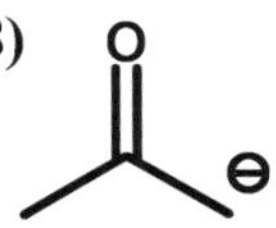

(C)

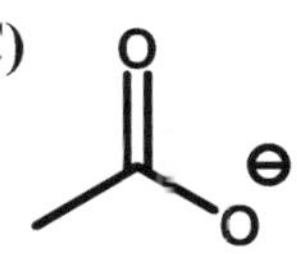

(D)

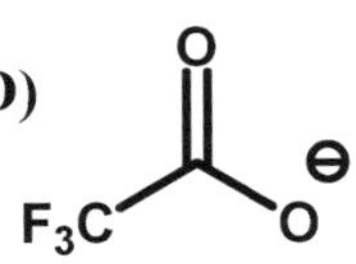

3. Which is the weakest base?

(A)

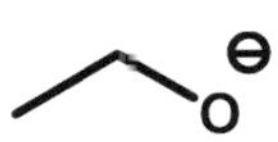

(B)

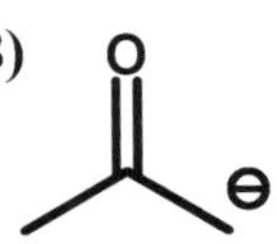

(C)

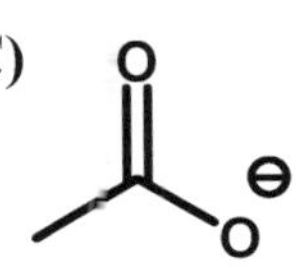

(D)

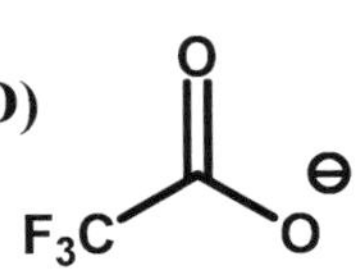

4. Which has the highest boiling point?

(A)

(B)

(C)

(D)

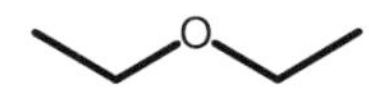

5. When 2-iodobutane is heated in the presence of water, the major product is:

(A) 1-butanol only

(B) 2-butanol only

(C) a mixture of 2-butanol and *trans*-2-butene

(D) No reaction

6. Provide the major product of the following E2 Reaction:

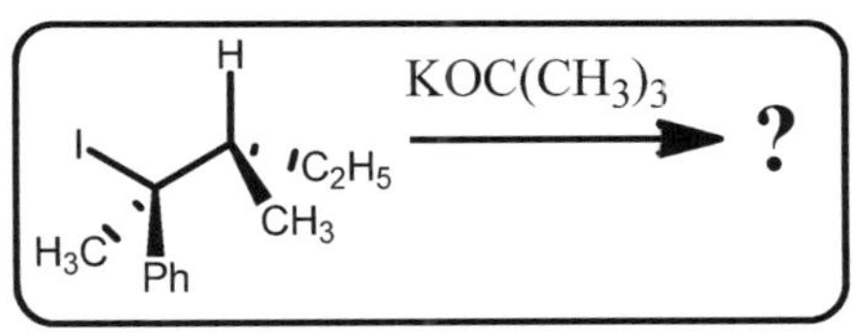

(A)

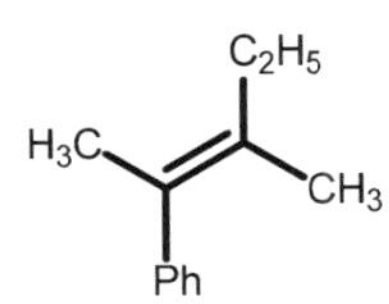

(B)

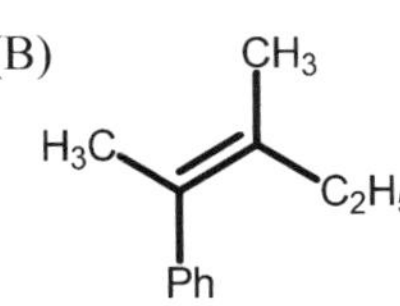

(C)

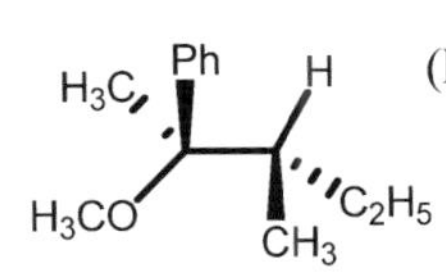

(D)

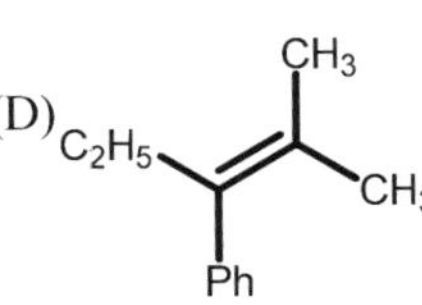

7. Which reaction coordinate diagram corresponds to the reaction sequence shown?

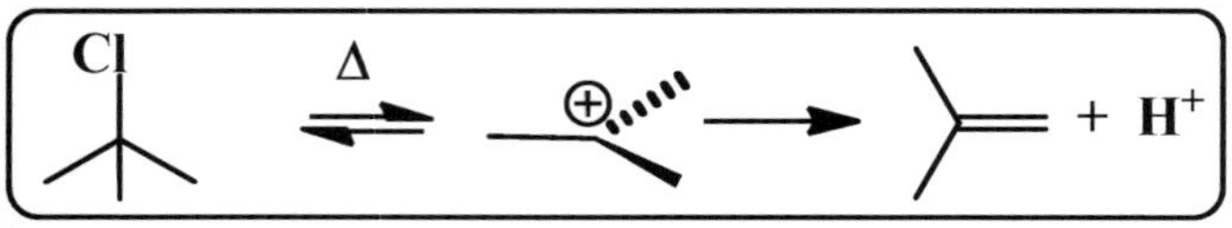

(A)

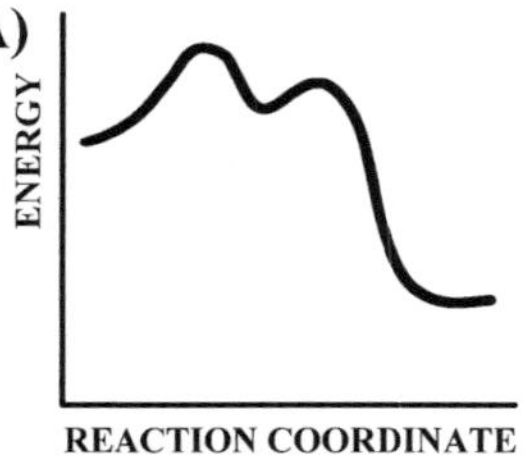

(B)

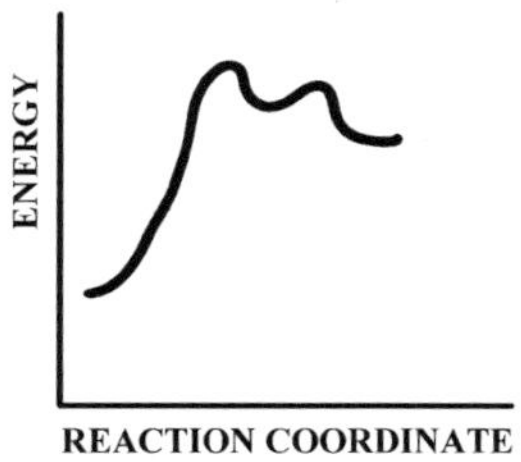

(C)

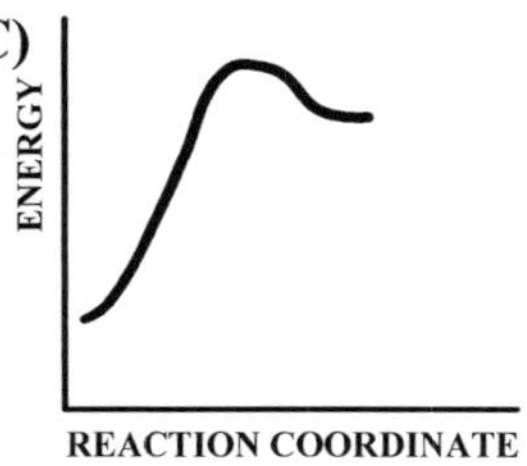

(D)

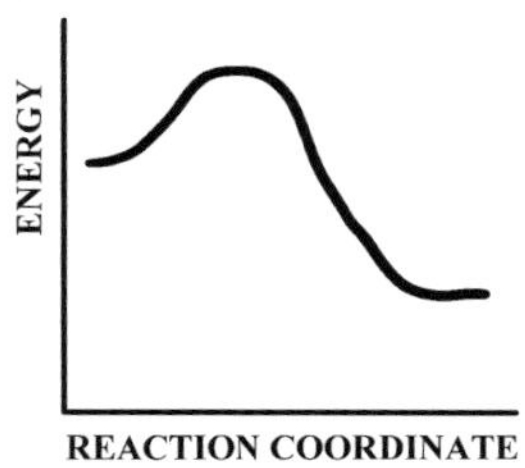

8. Which R-Br reacts fastest upon reaction with NaCN?

(A)

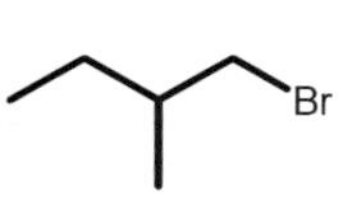

(B)

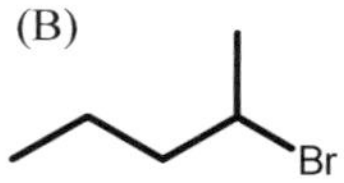

(C)

(D)

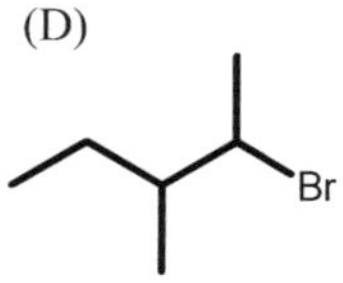

9. Which of the following reactions would give the best yield of 1-methylcyclopentanol?

(A)

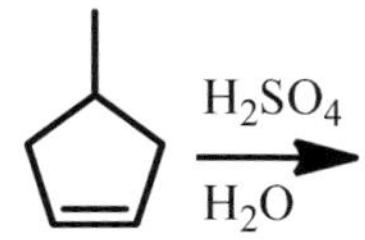

(B)

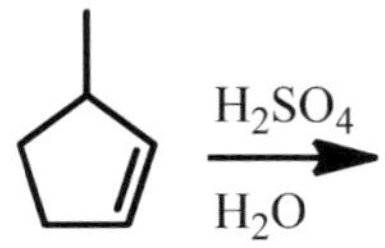

(C)

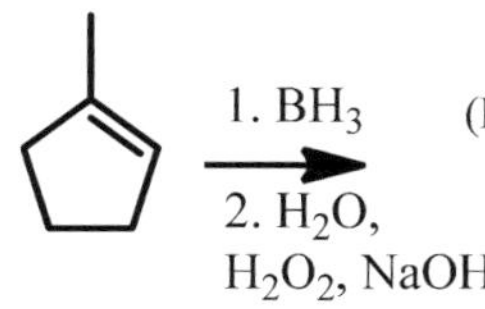

(D)

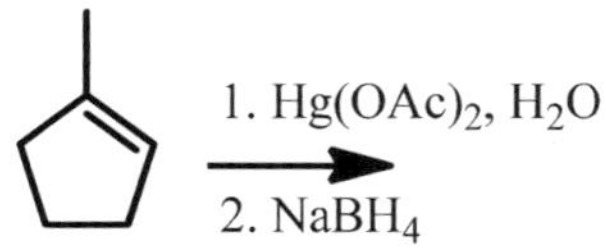

10. What is the product of the indicated reaction?

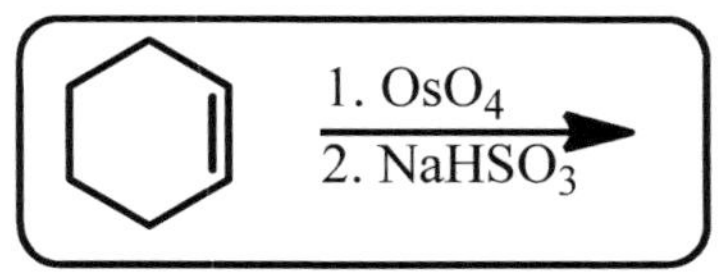

(A)

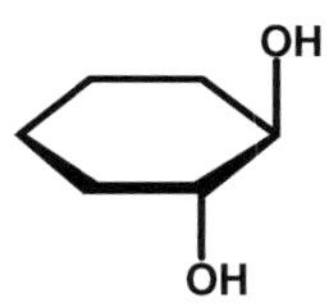

(B)

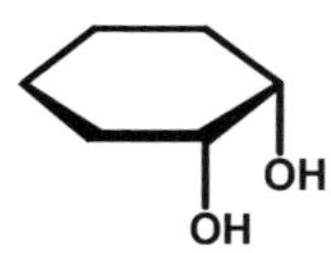

(C)

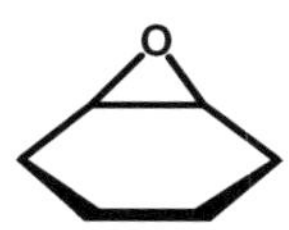

(D)

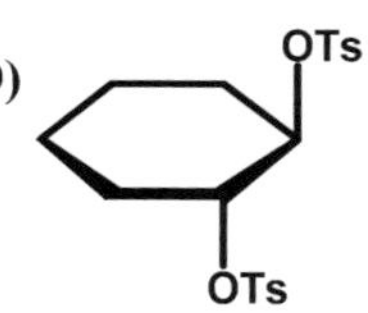

11. Which best represents the major product(s) of the indicated ozonolysis reaction?

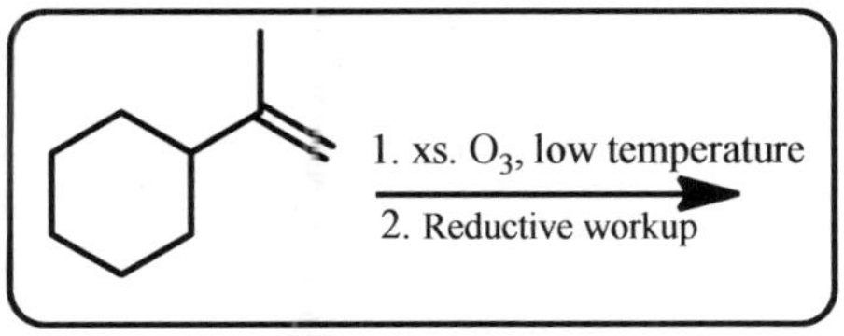

(A)

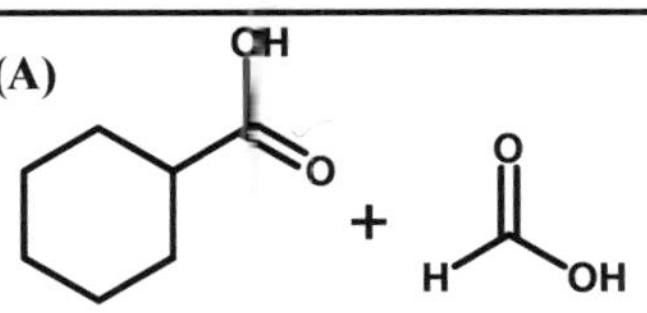

(B)

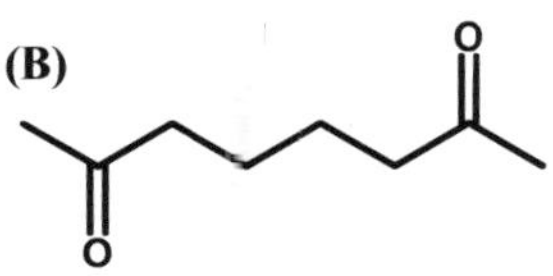

(C)

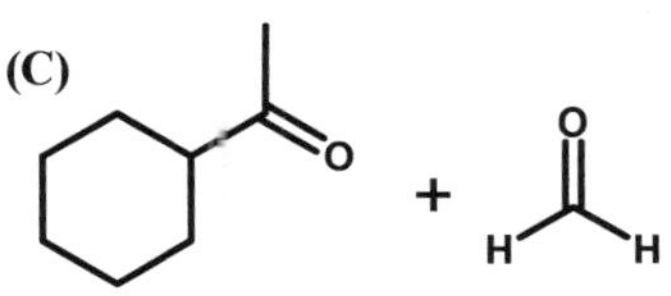

(D)

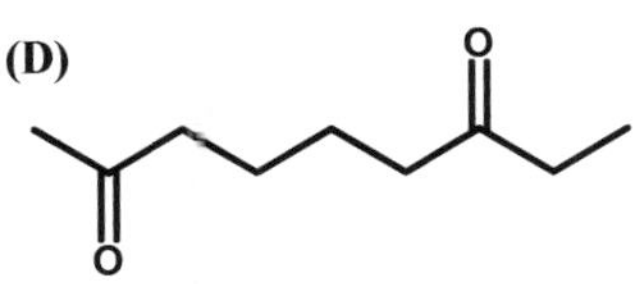

12. Which reaction is concerted?

A) 2-butene + H_2O/H_2SO_4
B) 2-butene + Br_2
C) 2-butene + HCl
D) 2-butene + peroxyacid
E) 2-butene + HBr /peroxides

13. In which solvent can one prepare a Grignard reagent without the solvent itself reacting with the Grignard reagent?

A) 1-butanol
B) $CH_3CH_2OCH_2CH_3$
C) $CH_3CH_2CH_2NH_2$
D) acetone (or 2-propoanone)
E) acetic acid

14. What is the major product of the reaction sequence?

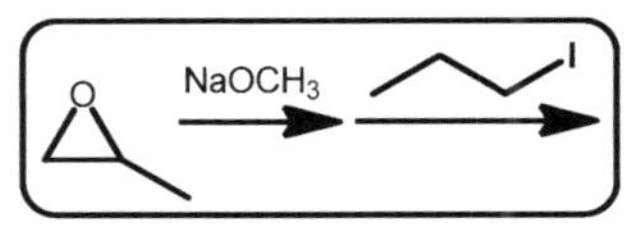

(A)

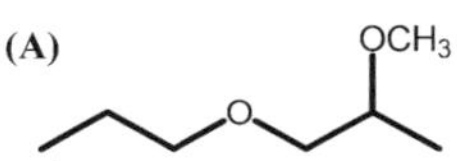

(B)

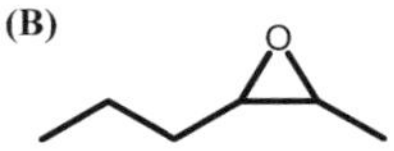

(C)

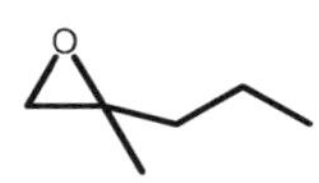

(D)

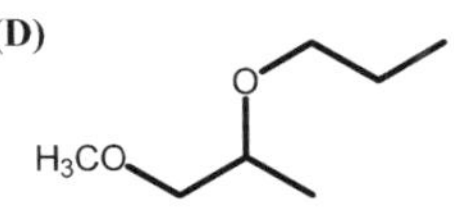

15. Which is the most reactive of these carbonyl-bearing functional groups with respect to nucleophilic attack?

(A) Carboxylate
(B) Acid chloride
(C) Amide
(D) Ester

16. Which R-Br reacts fastest in this S_N1 reaction?

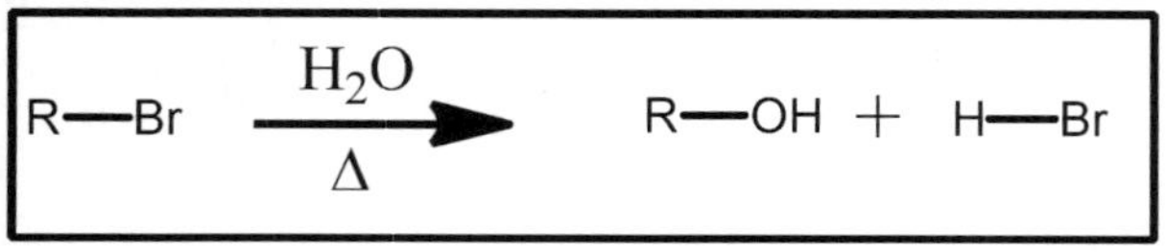

(A)

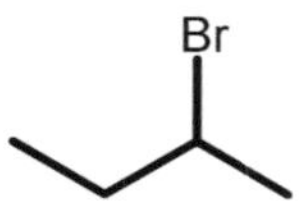

(B)

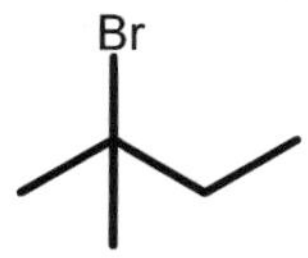

(C)

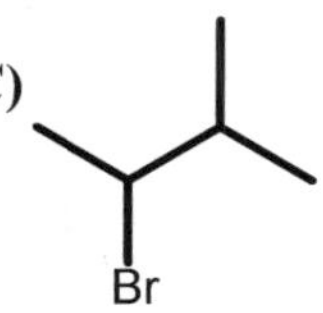

(D)

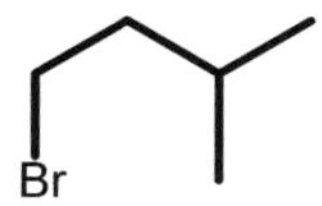

17. Which **site** will react fastest to nitration (or any electrophilic aromatic substitution) reaction?

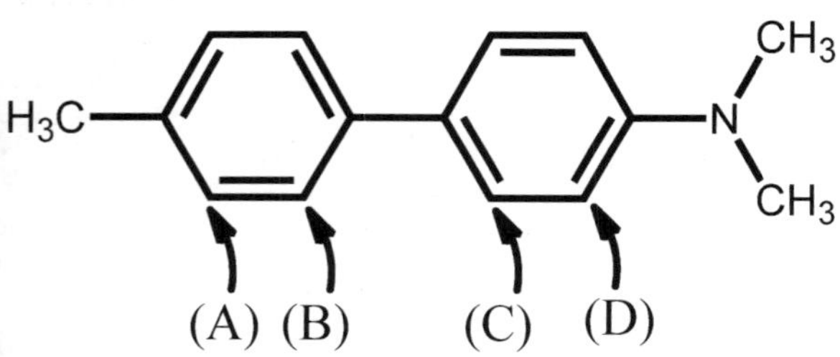

18. What is the major product of this reaction?

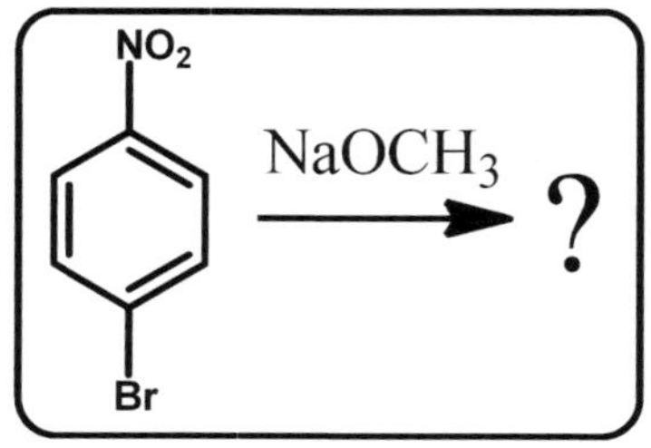

(A)

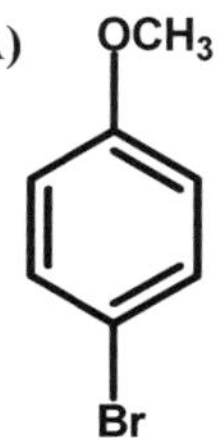

(B)

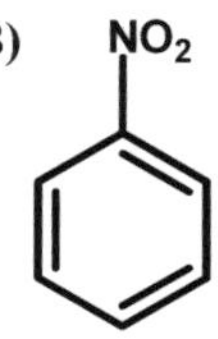

(C)

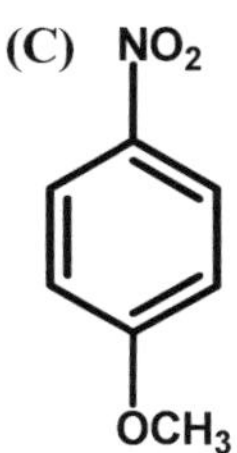

(D)

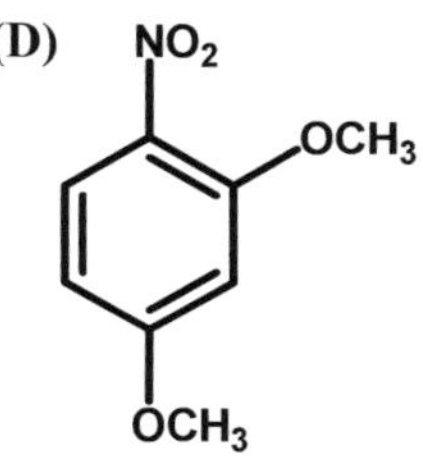

19. Which intermediate is formed when KNH_2 reacts with bromobenzene?

(A)

(B)

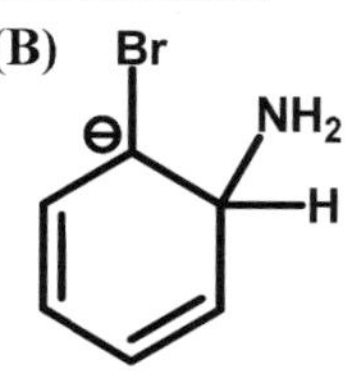

(C)

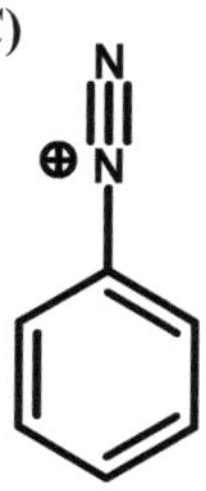

(D)

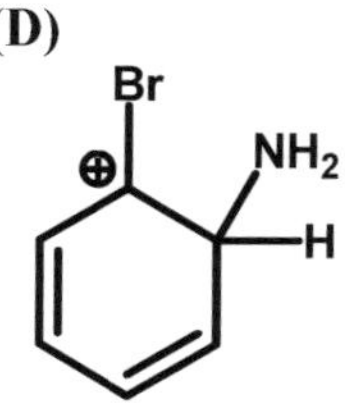

20. Which is/are aromatic?

(A) (B) (C)

(D) All are aromatic

(E) None are aromatic

21. Which is a major product of the indicated reaction?

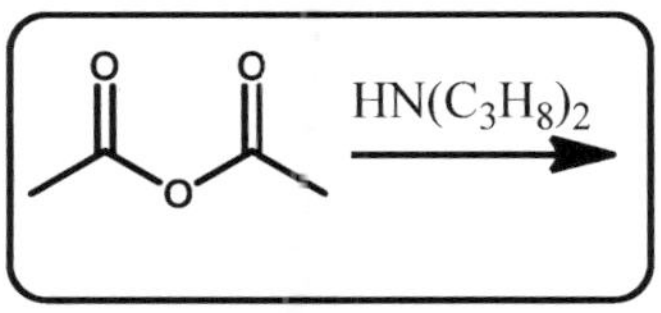

(A)

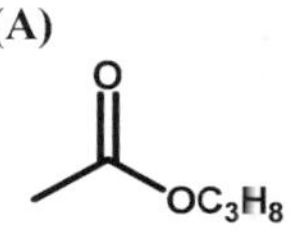

(B)

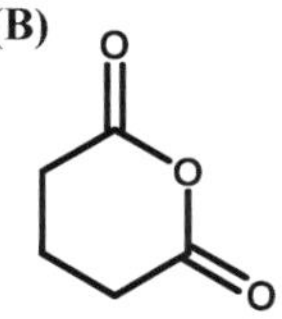

(C)

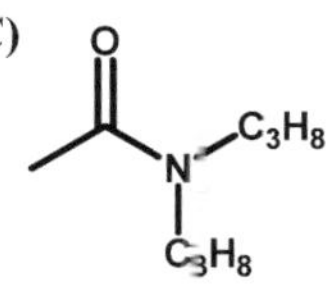

(D)

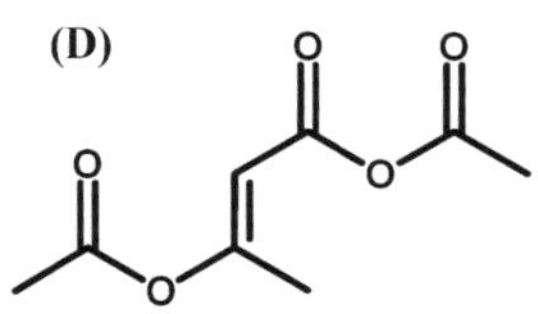

22. Which is the major product this reaction?

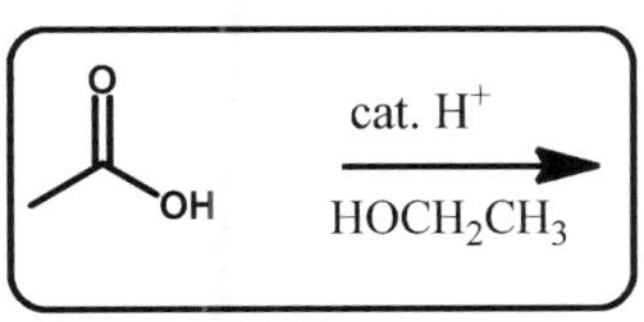

(A)

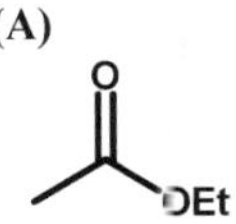

(B)

(C)

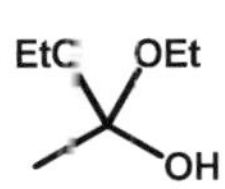

(D) No Net Reaction

23. Which is the major product of the indicated reaction?

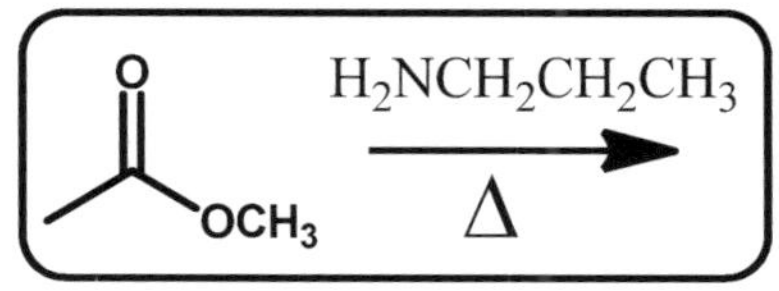

(A)

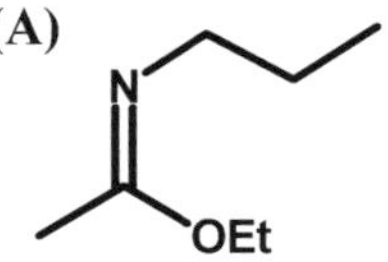

(B)

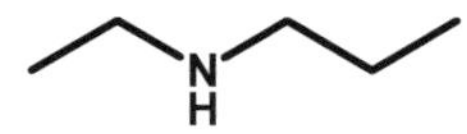

(C)

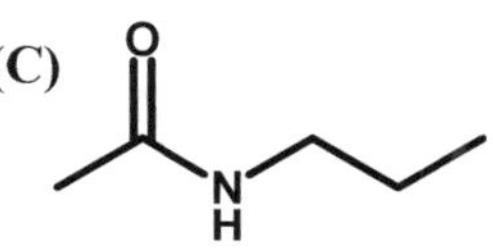

(D) No Net Reaction

24. Which is the major product of the indicated reaction?

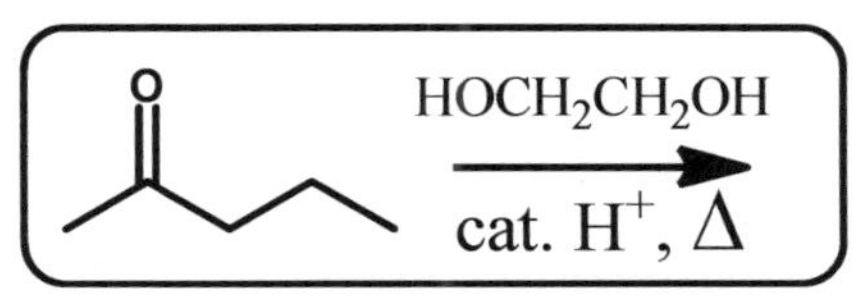

(A)

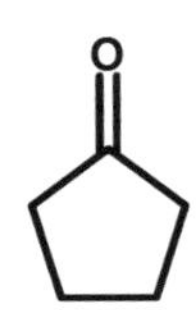

(B)

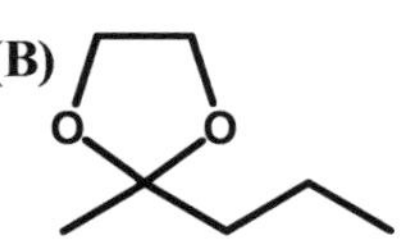

(C)

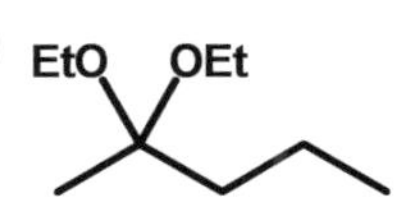

(D) No Net Reaction

25. Which starting material gives the indicated ROMP product shown in this reaction?

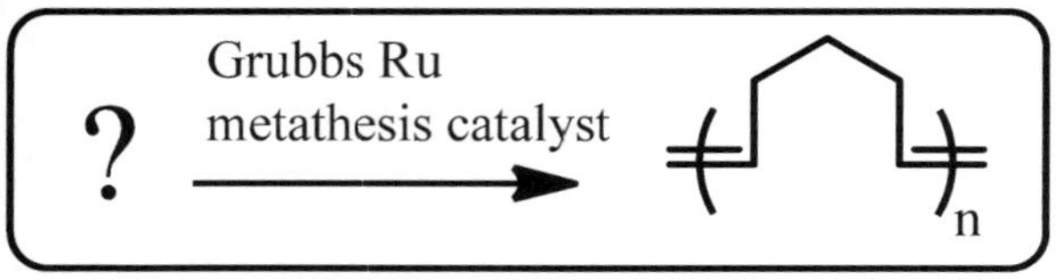

(A) 

(B) 

(C) 

(D) 

26. Which is the major product of the indicated reaction?

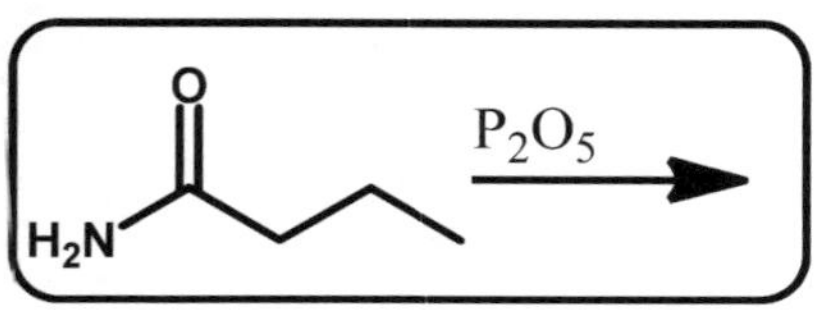

(A) 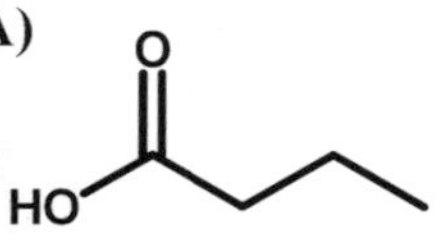

(B) 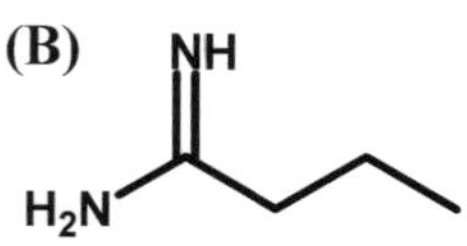

(C) 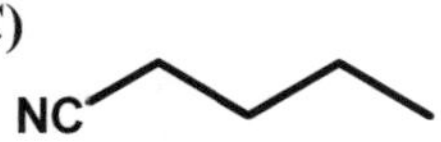

(D) 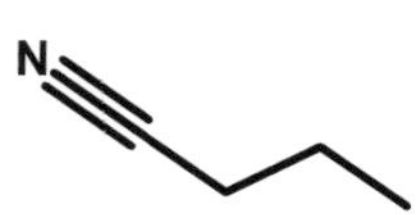

27. What is the major product of reaction if benzonitrile undergoes acid-catalyzed hydrolysis?

A) benzaldehyde
B) benzoic acid
C) aniline
D) nitrobenzene

28. Which reacts fastest to nucleophilic aromatic substitution?

(A) 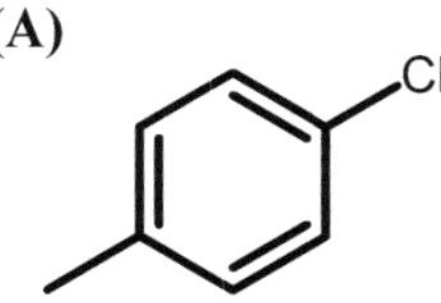

(B) 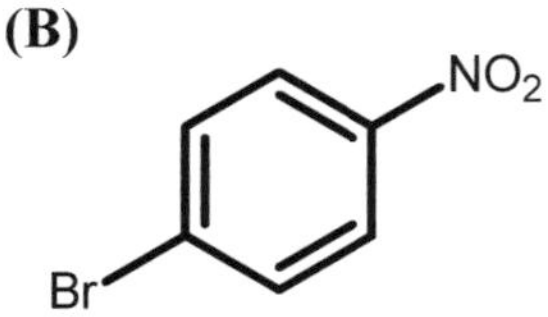

(C) 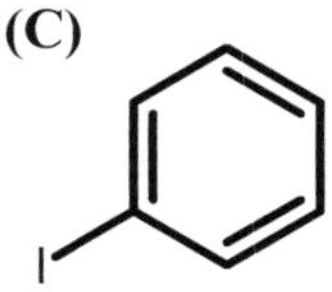

(D) 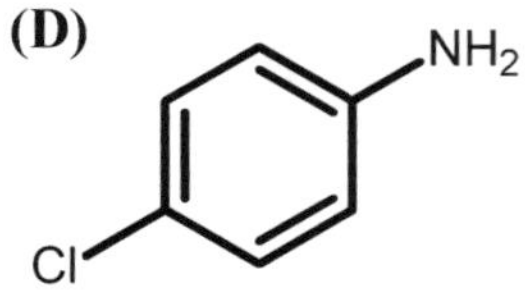

29. Which is the missing reagent needed for this reaction via a **Suzuki** coupling reaction?

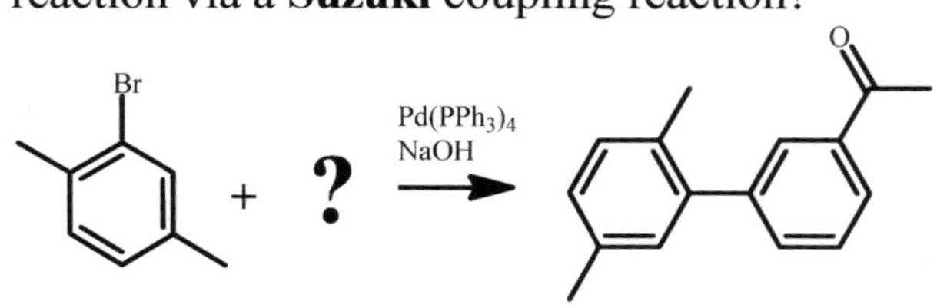

(A) 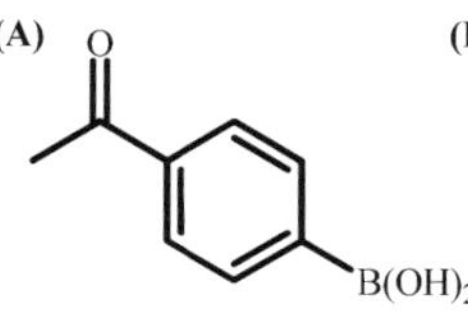

(B) 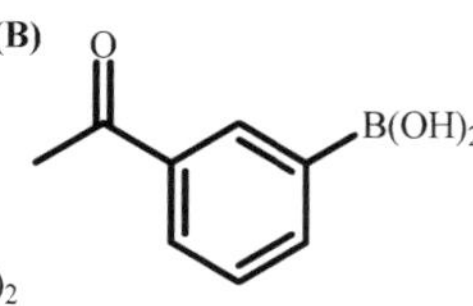

(C) 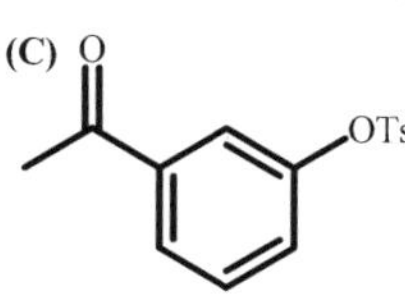

(D) 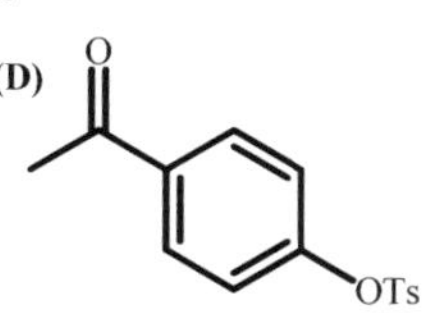

30. Which is a major product of this reaction sequence?

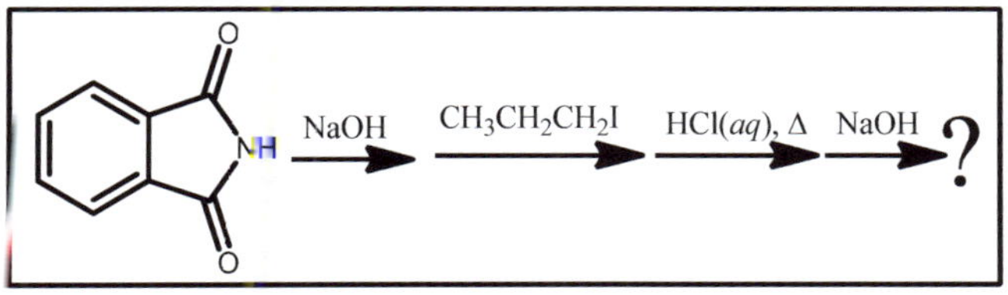

(A) *n*-Propyl phthalimide
(B) phenyl propyl amide
(C) phenyl propyl amine
(D) 1-propyl amine

Key for Progress Check 2:

1. D
2. A
3. D
4. B
5. C
6. B
7. A
8. C
9. D
10. B
11. C
12. D
13. B
14. D
15. B
16. B
17. D
18. C
19. A
20. D
21. C
22. A
23. C
24. B
25. B
26. D
27. B
28. B
29. B
30. D

Detailed Feedback for Progress Check 2

For each question that you missed or were unsure of, review the Necessary Knowledge and Reasoning and possible mistakes provided below. By addressing the gaps in your knowledge and avoiding similar mistakes on future problems, you should see improvement in your scores.

1. *Necessary Knowledge and Reasoning:*

a. The *strongest* acid is the one that produces the *most* stable anion upon removal of the proton.

b. A more electronegative element loses its proton more easily than a less electronegative element does. Applying that concept to this case, the H on the O is removed rather than any of the H atoms on carbons. You should thus compare the anions formed by taking a proton off of the O in each molecule.

c. If the negative charge can be delocalized via resonance, it acquires greater stability as compared to an anion that lacks resonance delocalization but has the negative charge on the same atom (here, oxygen). Only choice D has an anion that can be stabilized via resonance:

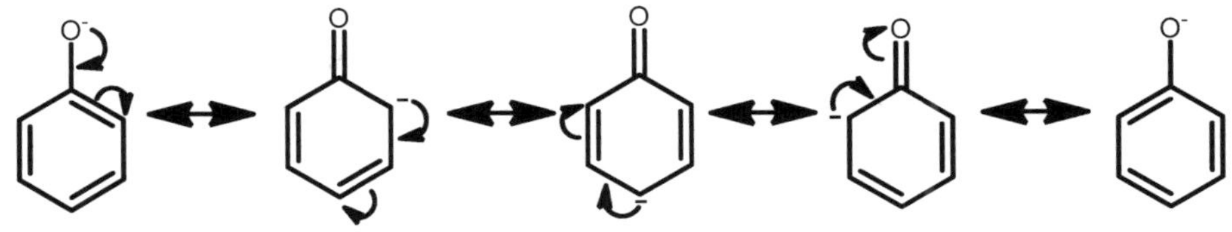

The answer is D.

2. *Necessary Knowledge and Reasoning:*

a. A nucleophile is something that donates electrons to an electron acceptor (like an electrophile).

b. Here, the anion is on oxygen in every choice. When comparing a set of nucleophiles, all of which have the negative charge on atoms in the same row, the nucleophilicity trend follows the same order as that for basicity. The stronger base will also be the better nucleophile.

c. The stronger base is the one that is the less stable anion.

d. It is important to consider resonance when evaluating stability; one cannot assume that the most stable resonance contributor will be the one that is shown in a problem.

e. We also must consider the electronegativity of the element on which the negative charge is positioned. The anion will be more stable when it is on a more electronegative element (provided the size of the elements is comparable, i.e., they are in the same row).

f. Choices B, C and D can all engage in resonance delocalization of the negative charge:

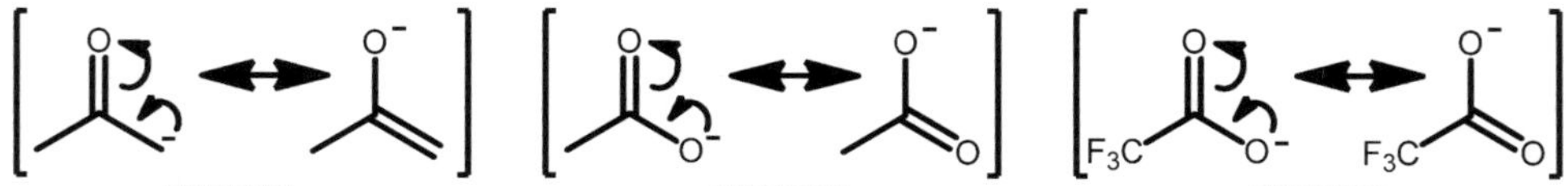

CHOICE B CHOICE C CHOICE D

All choices also have at least one resonance contributor in which the negative charge is on O.

g. If the anion can engage in resonance stabilization of the negative charge through delocalization, the anion will become more stable than an anion in which the negative charge is on the same element (like here, where the negative charge is on oxygen in at least one resonance contributor in every choice).

h. Choice A is the only one that cannot engage in resonance delocalization, so it is the least stable anion. This means that choice A is also the strongest base, and, in this case, that means it is also the best nucleophile.

The answer is A.

3. *Necessary Knowledge and Reasoning:*

a. The weakest base is the species that is the most stable anion.

b. We saw in the previous question that resonance delocalization in choices B, C and D makes them more stable than choice A, so we rule out choice A as being the *most* stable anion; choice A is the *least* stable anion and the strongest base.

c. Of the remaining choices, choices C and D each have two resonance contributors wherein the negative charge is on O in both contributors:

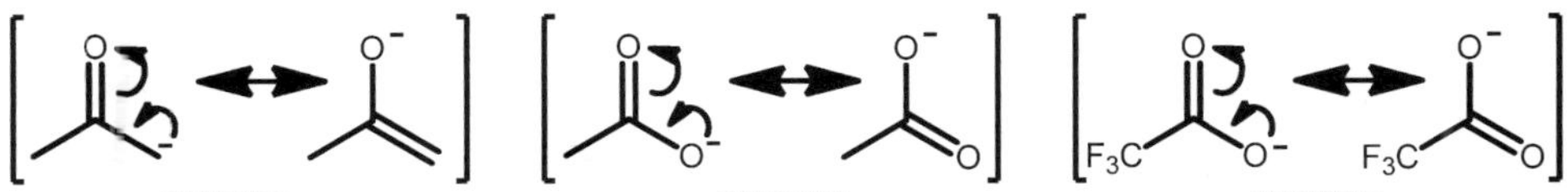

CHOICE B CHOICE C CHOICE D

In contrast, choice B features one resonance contributor in which the negative charge is on a carbon. Because a negative charge is more stable on a more electronegative element, this means that choice B is overall less stable than either choice C or D. Since we are looking for the most stable anion, we rule out choice B as a possible correct answer.

d. If an attractive force exists in a species, it helps to stabilize that species in comparison to an analogous species that lacks such an attractive force.

e. Because F is more electronegative than C, the C-F bonds in choice D are polar:

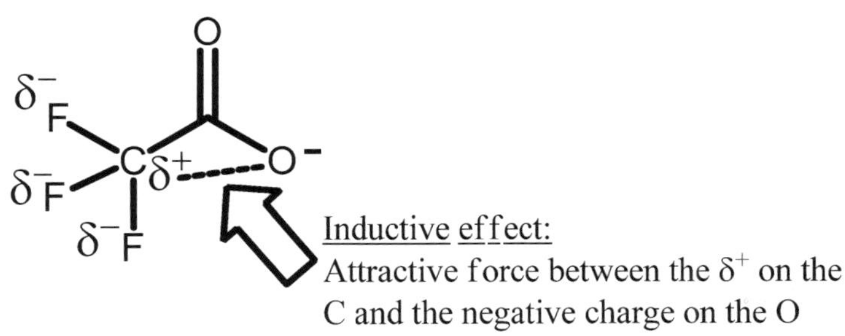

f. The partial positive charge on the C will attract the negative charge on the oxygen (indicated by the dashed line in the figure above; it is NOT a bond, just a force). This type of stabilizing interaction resulting from a charge induced by an electronegativity difference is called an **inductive effect**. The attractive inductive force in choice D leads to D being a more stable anion than C.

The answer is D.

4. *Necessary Knowledge and Reasoning:*

a. The boiling point is the temperature at which a liquid becomes a gas.

b. Molecules in the gaseous phase are farther apart than are molecules in the liquid phase. In fact, the intermolecular forces between gas-phase molecules are negligible because the molecules are so far apart.

c. The stronger the intermolecular forces between molecules in a sample of a liquid, the more thermal energy (higher temperature) will be needed to break them apart into a gaseous state.

d. Hydrogen bonding is the strongest intermolecular force present between neutral organic molecules. Hydrogen bonding is possible in molecules that have H-O, H-N, or H-F bonds. Choices A and B are thus the only choices capable of hydrogen bonding.

e. Because B has two O-H units in it, whereas A has only one, the intermolecular forces between molecules of B are stronger. This means that B will have the highest boiling point of all choices.

The answer is B.

5. *Necessary Knowledge and Reasoning:*

a. The structure of 2-iodobutane is:

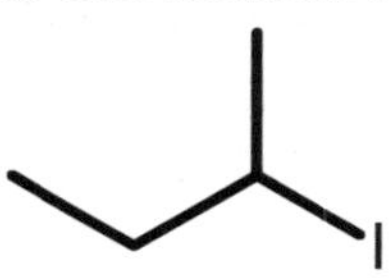

b. Water is neither a good nucleophile nor a strong base. It is a weak base and a poor nucleophile.

c. When a molecule with a good leaving group (here iodine) is heated in the absence of a good nucleophile/strong base, thermal energy will break the weakest bond in the molecule if a high enough temperature is reached. In this case, heterolytic cleavage of the C-I bond occurs to make a carbocation:

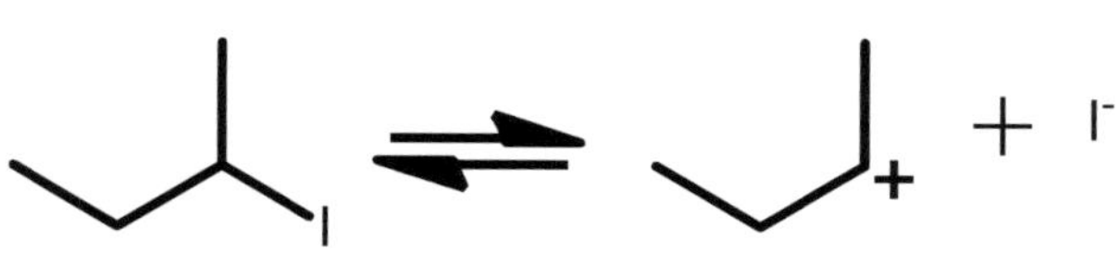

d. Generally, this process is very unfavorable if the cation formed is a primary carbocation. Here, however, the carbocation is secondary. A secondary carbocation has enough stability for the reaction to proceed.

e. Once a carbocation forms, there are three fates it can undergo.

i. One fate is carbocation rearrangement. Rearrangement only happens if rearrangement would lead to the formation of a more stable species. There are no such possible pathways available to the carbocation formed, so we rule out rearrangement.

ii. The second possible fate is the coordination of water (with simultaneous loss of proton from the water) to form an alcohol. This is leads to the S_N1 product and will be one major pathway available in the current example:

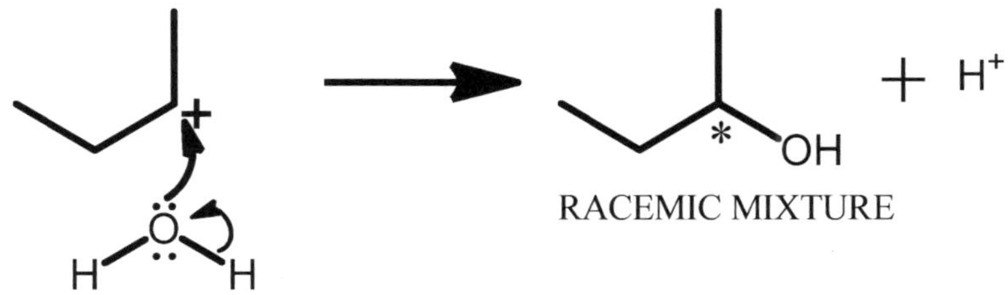

iii. The third possible fate is electrophilic elimination of a proton to form a pi bond. This would be an E1 product and is the other major reaction pathway available in the current case. In order to select the H to be removed, we have to remember that when forming an alkene in a reaction. The more substituted alkene is favored because the more substituted alkene is derived via the lowest energy transition state and is the more stable product. This observation is essentially "**Zaitsev's rule**": The more stable alkene is the major product. Furthermore, if a disubstituted alkene is made, the *trans*-isomer will be formed where the mechanism allows. The E1 product in the current case is thus *trans*-2-butene:

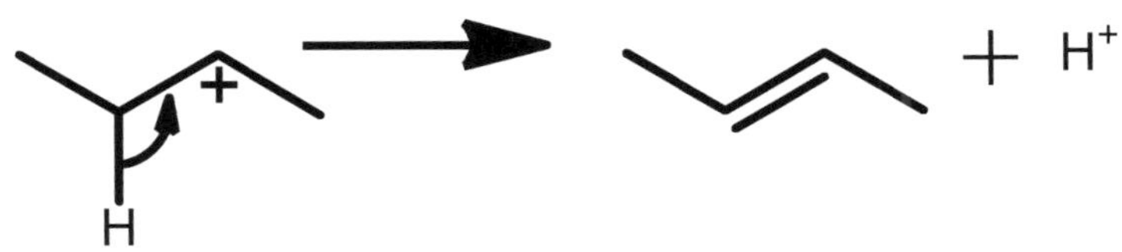

f. The major product is thus a mixture of both the SN1 product (2-butanol) and the E1 product (*trans*-2-butene).

The answer is C.

6. *Necessary Knowledge and Reasoning:*

a. An E2 reaction is a concerted reaction, meaning it happens all at once without any intermediates that can rearrange.

b. In the E2 reaction, a pi bond forms between two carbons, one of which loses a hydrogen and the other of which loses a "leaving group" (group that can leave the molecule as a stable anion). Because choices C and D lack the pi bond produced by an E2 reaction, neither of these can be the correct answer.

c. The H and leaving group must be antiperiplanar to one another in the starting material. Antiperiplanar means that they must be in the same plane and anti to one another.

d. When forming an alkene in a reaction, like we do here, the more substituted alkene tends to form because the more substituted alkene forms via the lowest energy transition state and is the more stable product. This observation is essentially "**Zaitsev's rule**": The more stable alkene is the major product.

e. The alkene starting material has a H that can be removed from the most substituted carbon that is adjacent to the carbon with the leaving group (I) on it:

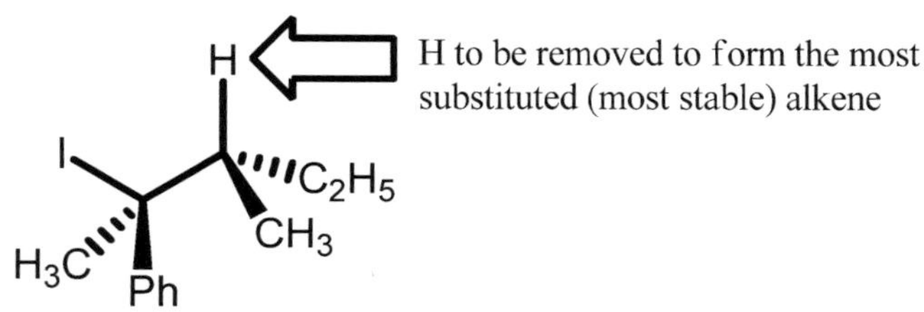

f. In order to line the molecule up properly to do an E2 reaction, we need to rotate around the C2-C3 bond to get the I and to-be-removed H anti to one another:

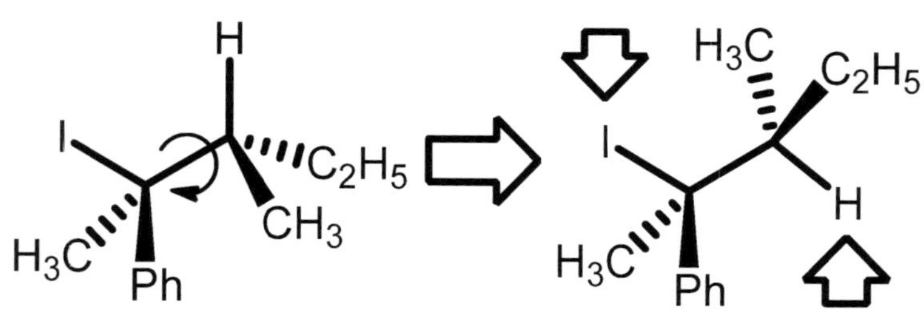

In the structure on the right, rotation around the indicated single bond has put the two necessary groups anti to one another (the two groups are marked by arrows in the structure on the right).

g. Now that the groups to be removed are antiperiplanar, an E2 reaction can take place.

h. Because the E2 reaction is concerted, all of the groups that are not removed from the molecule will remain in the same relative position as in the starting material. In this case, this means that the two groups that are both pointing towards us (the Ph and the C_2H_5 attached via wedge-shaped bonds) will remain pointing in the same direction in the product as well. Likewise, the two groups that both point away from us (the two CH_3 groups attached via hatched lines) will also remain pointing in the same direction in the product. Only choice B meets these criteria.

The answer is B.

7. *Necessary Knowledge and Reasoning:*

a. This is an example of an E1 reaction.

b. A carbocation intermediate is involved in this reaction, and intermediates are at local minima on a reaction coordinate diagram.

c. There will need to be an initial expenditure of energy (positive Y axis direction indicates energy expenditure) to get to the intermediate carbocation.

d. The lower on the Y axis a species is, the more stable it is.

e. The reaction would proceed poorly under these conditions if the products were less stable than the reactants.

f. Choice A is the only one that features an intermediate "valley" for the carbocation AND has lower energy products than reactants.

The answer is A.

8. *Necessary Knowledge and Reasoning:*

a. An ionic compound such as NaCN will dissociate to ions to some extent in polar solvents. In this case, Na^+ and ^-CN (cyanide ion) will be formed in solution.

b. Cyanide anion is a good nucleophile but a weak base.

c. When a good nucleophile that is also a weak base reacts with an alkyl halide, the most likely reaction pathway is the S_N2 reaction.

d. The S_N2 reaction gets slower the more steric bulk is surrounding the carbon to which the leaving group is attached. Therefore, the first thing to look for is whether the leaving group-bearing C is primary, secondary or tertiary. The rate of reaction will be: primary > secondary > tertiary. Here, this lets us rule out choices B and D as fastest.

e. Between A and B, both of which are primary alkyl halides, we note that there is a methyl group on the C neighboring the C to which the Br is attached. Such a substituent coming off the parent chain in the carbon right next to the site to which the leaving group is attached is called a β-branch. The presence of this β-branch makes substrate A somewhat more sterically crowded near where the nucleophile needs to attack, so S_N2 reaction of A will be slower than that of C.

The answer is C.

9. *Necessary Knowledge and Reasoning:*

a. Choice A is a hydration reaction.

i. A hydration reaction proceeds via a carbocation intermediate formed by protonation of the alkene.

ii. When an alkene is protonated, the more stable (more substituted) carbocation forms preferentially. Here, both sides of the alkene are equally substituted (both sides of the C=C bond are secondary carbons) and the molecule is symmetric, so we only get one possible carbocation:

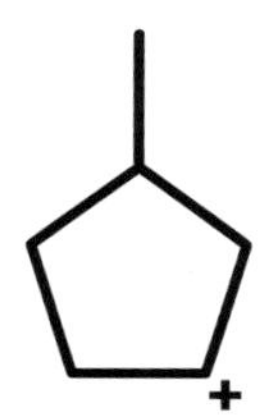

iii. This carbocation will not rearrange, so the OH will be added to the carbocationic carbon to form the final product, 3-methylcyclohexanol:

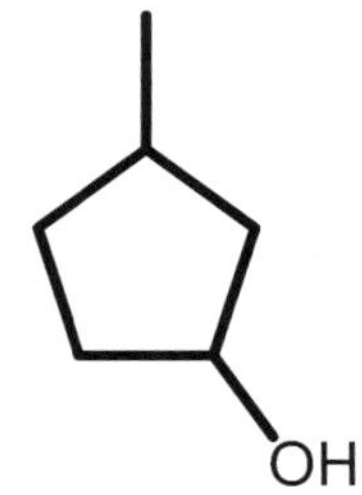

This is not the product that this question asks us to make, so we rule it out as the correct answer.

b. Choice B is also a hydration reaction.

i. As was seen for reaction A, both the C atoms in the C=C bond are secondary, so they are equally likely to end up as the cationic carbon in this reaction. Unlike reaction A, however, our reactant is not a symmetric molecule, so we have two possible carbocations formed in about 50% yield each:

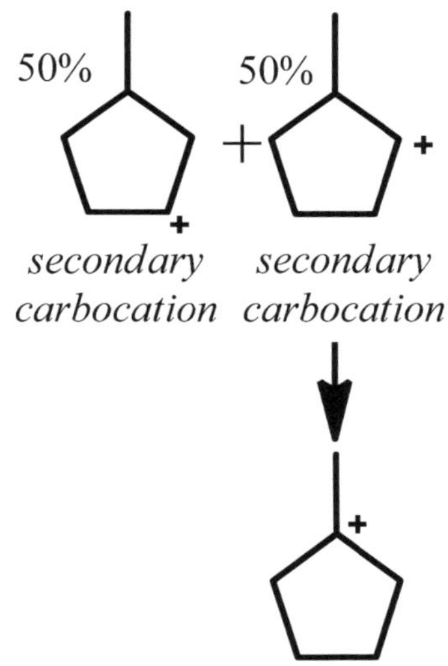

tertiary carbocation

50% of original material

ii. Notice, however, that one of the two originally-formed secondary carbocations will rearrange to a tertiary carbocation. We assume that nearly all of the cation on the upper right, above, will end up as the tertiary carbocation. This means that, of the original material, about half is the secondary cation (upper left, above), and half is the tertiary, rearranged cation:

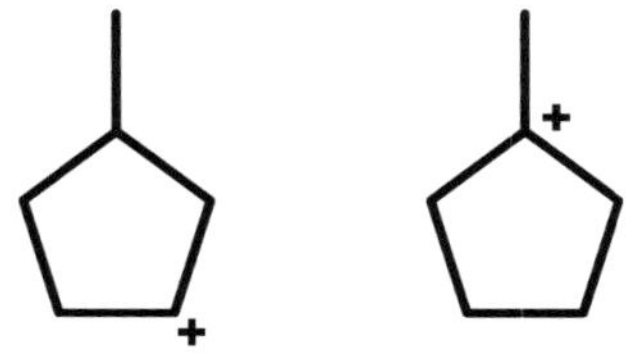

50% of original material 50% of original material

iii. Now that we have determined what carbocations we will have after all rearranging is done, we can see that the OH will add to the cationic sites. This will produce a roughly 50:50 mixture of two alcohols:

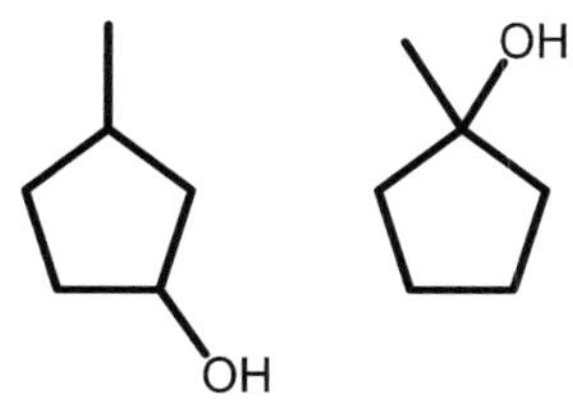

~50% yield ~50% yield

One of the alcohols is the target product, 1-methylcyclohexanol. The other is not the target. So, we have ~50% yield of the target. The question asked which reaction choice will give the *best* yield of the target 1-methylcyclohexanol, so we will still need to check the other answers to see whether any of those have a >50% yield.

C. Choice C is a hydroboration/oxidation sequence.

i. The hydroboration/oxidation is a non-Markovnikov addition where a hydrogen and an –OH add to the two carbons in the C=C bond with no rearrangement.

ii. In a non-Markovnikov addition, the *less* substituted carbon ends up making a bond to the *more* electronegative element added.

iii. This non-Markovnikov reaction thus forms 2-methylcyclohexanol, which is not the target product. We thus rule out choice C as a possible answer.

D. Choice D is an oxymercuration/reduction sequence where an H and an –OH add to the two carbons in the C=C bond with no rearrangement.

i. The oxymercuration/reduction sequence is a Markovnikov addition.

ii. In a Markovnikov addition, the *more* substituted carbon of the C=C bond ends up attached to the *more* electronegative element added.

iii. Here, this reaction leads to 1-methylcyclohexanol as the major product (>50%). This makes choice D a better choice than choice B, which gives only ~50% of the target.

The answer is D.

10. *Necessary Knowledge and Reasoning:*

a. On the basis of the reaction conditions, we recognize that this is an oxidation of an alkene to a vicinal diol.

b. A vicinal diol is one in which there are two alcohols on adjacent carbons:

General structures of a vicinal diol

Note that one can draw a vicinal diol with the two OH units pointed either the same direction or opposite directions. Only product choices A and B are vicinal diols.

c. The mechanism of OsO_4-mediated oxidation leads to a *syn*-addition of the two OH groups. Choice A is an *anti*-vicinal diol (the two OH units point in opposite directions). This leaves us with the only *syn*-vicinal diol (two OH units point in the same direction), choice B.

The answer is B.

11. *Necessary Knowledge and Reasoning:*

a. An ozonolysis reaction will split a C=C double bond and place a double bond to O on each of the carbons that used to be in the C=C double bond:

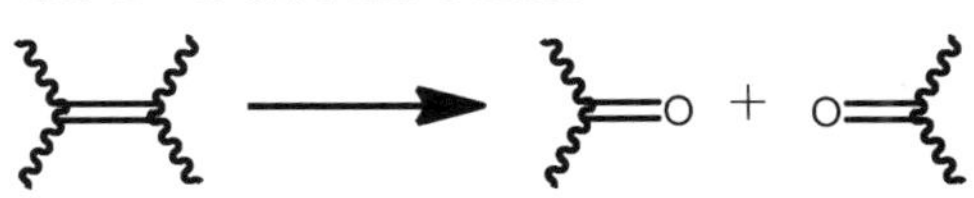

b. There are two kinds of workups for ozonolysis:

i. Oxidative workup (usually with H_2O_2) makes any H on an sp^2 carbon into an OH in the product. The H atoms that change into OH units are circled in the figure below:

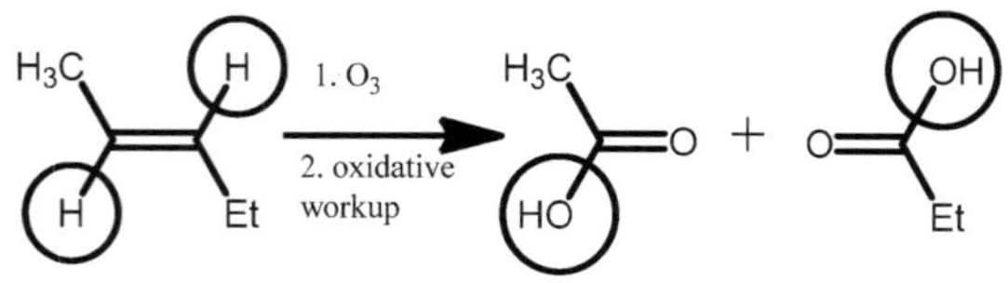

ii. Reductive workup (usually with Zn/H^+ or SR_2) leaves all of the H atoms as they were.

b. Here, we need to cut the C=C bond, adding a double bond to O to each carbon that was in the C=C bond and leave the H atoms unchanged:

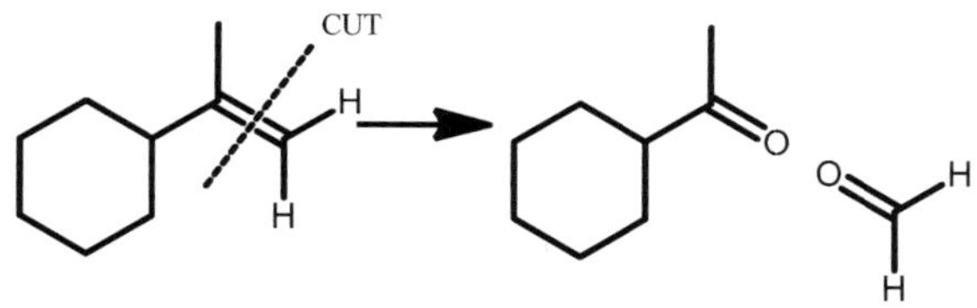

This set of products matches that shown in choice C.

The answer is C.

12. *Necessary Knowledge and Reasoning:*

a. The concerted reactions generally taught up to this point in introductory organic chemistry courses are: E2 reaction, S_N2 reaction, Cyclopropanation, epoxidation and Diels-Alder reaction. Note that although the hydroboration part of the hydroboration/oxidation sequence is concerted, the entire sequence involves several steps.

b. Choice D is the only reaction that is in our list of concerted reactions (it is an epoxidation).

The answer is D.

13. *Necessary Knowledge and Reasoning:*

a. A Grignard reagent is of the form XMgR, where X is a halogen (Cl, Br or I).

b. For the purpose of use in mechanisms, one can often think of Grignard reagents as ionic compounds that form carbanions in solutions. They do not actually dissociate into isolated ions, but the reactivity often reflects this estimation:

$$R_3C{-}Mg{-}X \rightleftharpoons R_3C^- + Mg^+{-}X$$

c. A carbanion is generally a very strong base.

d. If a strong base comes in contact with a proton that it can take to produce a more stable anion, this will be a favorable reaction.

e. An anion with a negative charge on a more electronegative atom will be more stable than an anion with a negative charge on a less electronegative element of about the same size (same row).

f. The pieces of information above lead us to conclude that any molecule that has a hydrogen on an O or N will have the O or N deprotonated by a Grignard to form a more stable anion. This means that choices A, C and E would be deprotonated by a Grignard:

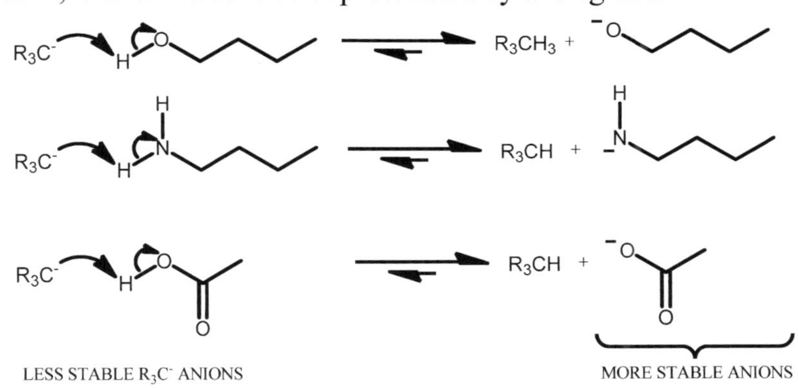

These three reactions are favored in the forward direction because the species on the right are MUCH more stable than the species on the left.

g. Grignard reagents are also typically very good nucleophiles. An exception might be a very bulky Grignard reagent, but this question asks for the ability to make any Grignard reagent in general, so we do not need to find specific exceptions. Because Grignard reagents can be very good nucleophiles, we cannot use any solvents that react with nucleophiles.

h. Carbonyls are susceptible to nucleophilic addition reactions. Choice D is a carbonyl, so it will undergo nucleophilic addition with Grignard reagents:

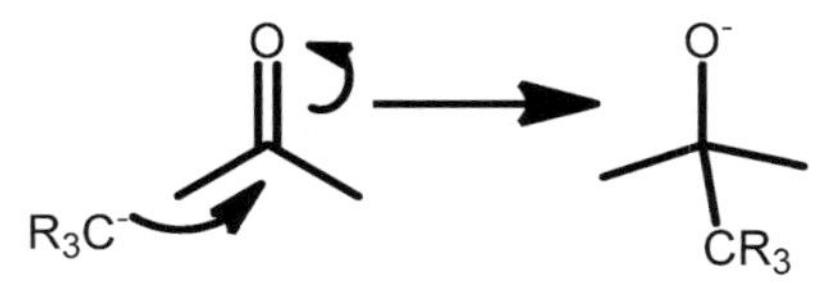

i. This leaves us with choice B as the only solvent that will not be expected to react with a Grignard reagent.

The answer is B.

14. *Necessary Knowledge and Reasoning:*

a. When a nucleophile reacts with an epoxide under **acidic conditions**, the nucleophile will attack the **more substituted** carbon of the epoxide ring. When a nucleophile reacts with an epoxide under basic conditions, the nucleophile will attack the **less substituted** carbon of the epoxide ring.

b. The first reactant in this problem is $NaOCH_3$. This is an ionic compound, so it will dissociate into Na+ and $^-OCH_3$ ions in solution.

c. The methoxide ($^-OCH_3$) ion is a strong base. Because a strong base is present, this nucleophile will attack the less substituted carbon of the epoxide ring:

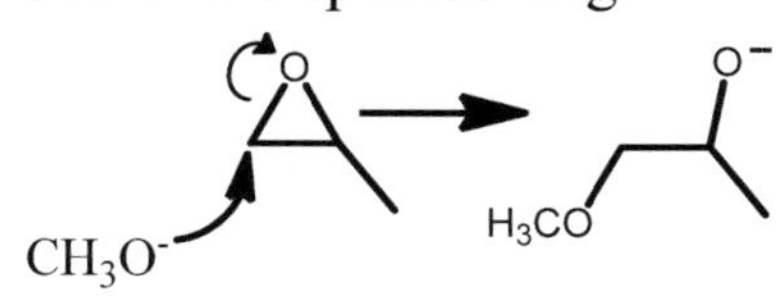

This is an example of an S_N2 reaction.

d. Now we are left with an anion, which will then react with the next reactant, 1-iodopropane. The anion is both a strong base and a good nucleophile. When a species like this reacts with a primary alkyl halide (with halogen = Cl, Br or I), an S_N2 reaction will occur:

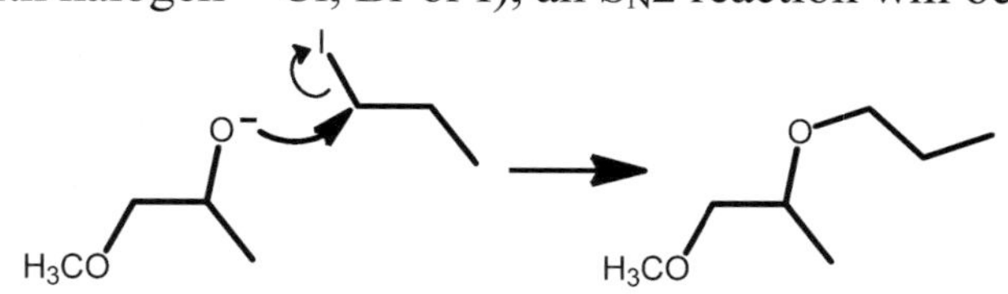

The answer is D.

15. *Necessary Knowledge and Reasoning:*

a. A nucleophile will more-readily attack a site that

i. has a higher partial positive charge on it,

ii. is less sterically hindered, and

iii. does not exhibit Coulombic repulsion (like charges repelling each other).

b. In order to evaluate the above criteria for the answer choices, we have to consider any possible resonance structures, the charge on the molecule and the polarity of the bonds:

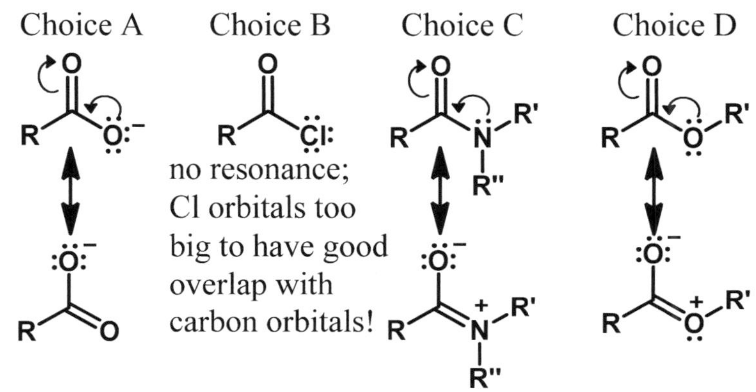

c. For choice A, any nucleophile that approaches the carbonyl carbon will experience Coulombic repulsion: negative charge repels negative charge.

d. In choices C and D, there are lone pairs on an atom that is directly attached to the carbonyl carbon. This situation will lead to donation of electrons from the adjacent atom to the carbonyl carbon. This donation of electrons to the carbonyl carbon will make it less susceptible to nucleophilic attack.

e. Choice B, an acid chloride also has lone pairs on an atom (chlorine) that is directly attached to the carbonyl carbon. However, because Cl is in the third row of the periodic table, it has much bigger valence orbitals than carbon. Orbitals that are different sizes do not overlap well with one another to form pi bonds, so chlorine will not engage in resonance via donation of its lone pair to carbon.

The answer is B.

16. *Necessary Knowledge and Reasoning:*

a. The S_N1 reaction mechanism involves a carbocation intermediate.

b. A carbocation is stabilized via *hyperconjugation* with any adjacent hydrocarbon substituents, so the stability of carbocations follows the general trend:

tertiary > secondary > primary > methyl

c. The more stable the carbocation intermediate, the faster the reaction.

The answer is B.

17. *Necessary Knowledge and Reasoning:*

a. An electrophilic aromatic substitution involves a carbocation intermediate.

b. A carbocation is stabilized by inductive electron donation and by resonance electron donation.

c. Resonance electron donors stabilize cations more than do inductive donors.

d. The more stable the cation intermediate, the faster the reaction.

e. Hydrocarbon substituents are inductive electron donors; they can stabilize cations via hyperconjugation.

f. The $–N(CH_3)_2$ unit is a resonance electron donor; the lone pair on N can be donated to a cation on the ring to which the $–N(CH_3)_2$ unit is attached.

g. On the basis of the facts discussed in points a-f (above), we conclude that a ring with a resonance donor on it will react faster than will a ring with an inductive donor on it. We therefore focus on choices C and D.

h. Resonance donors are *ortho/para* directing. This is because substitution at the *ortho* or *para* positions creates a carbocation that can be stabilized by resonance with contribution of the N lone pair. Choice C is *meta* to the $–N(CH_3)_2$ unit, whereas choice D is *ortho* to the $–N(CH_3)_2$ unit. This makes site D the fastest-reacting site.

The answer is D.

18. *Necessary Knowledge and Reasoning:*

a. The reactant, $NaOCH_3$, is a strong base and a good nucleophile.

b. The benzene ring has a strong resonance withdrawing group ($-NO_2$) and a good leaving group (Br) as substituents.

c. The leaving group (Br) is *para* to the withdrawing nitro group.

d. The presence of a good nucleophile and an aromatic ring with BOTH an electron withdrawing group and a good leaving group that is either *ortho* or *para* to the withdrawing group is the set of conditions necessary for a nucleophilic aromatic substitution reaction.

e. In a nucleophilic aromatic substitution reaction, the nucleophile replaces the leaving group. In this case, it means that a methoxy group ($-OCH_3$) will replace Br. Choice C shows this product.

The answer is C.

19. *Necessary Knowledge and Reasoning:*

a. The reactant, KNH_2, is a strong base and a good nucleophile.

b. The benzene ring has only a bromine on it.

c. We can rule out nucleophilic aromatic substitution because that reaction requires: i) strongly electron withdrawing group on the ring AND ii) a leaving group that is *o*- or *p*- to the withdrawing group.

d. The strongly basic KNH_2 will thus remove the most acidic proton on the bromobenzene.

e. The most acidic H is the one *ortho*- to the Br because it will be stabilized by the inductive effect of the Br.

f. Once the proton is removed, a carbanion results (left side of figure below in point g).

g. Because carbanions are unstable compared to bromide, it is a favorable reaction to push the negative charge onto the bromide:

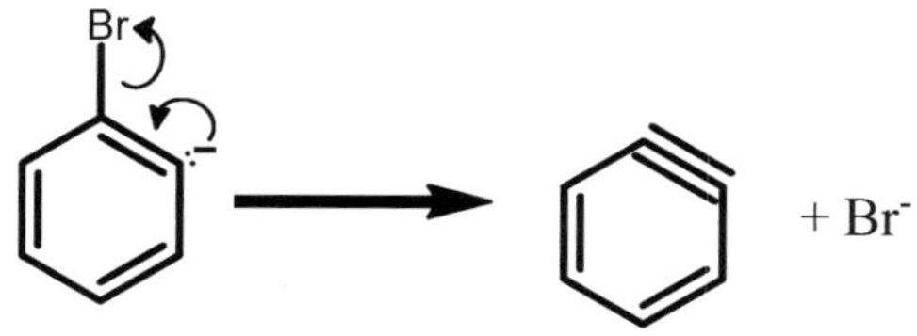

Benzyne

This intermediate matches choice A.

The answer is A.

20. *Necessary Knowledge and Reasoning:*

a. The criteria for aromaticity are:

i. Must be cyclic and planar.

ii. Must contain a continuous pi conjugated system around the entire ring (you can usually identify this easily by seeing if a pi bond can go to every atom in the ring in at least one of the resonance structures of the compound).

iii. Must contain 4n+2 electrons in the pi system, where n is any integer. This is the Hückel rule.

b. When an atom in the potentially aromatic ring has a lone pair on it, it will donate the lone pair to the pi system if doing so will make the system aromatic. It is important to note that only one lone pair can be donated into a pi system by any given atom in the ring.

c. If we look at the choices and keep the above points in mind, we see that A, B and C can all be aromatic because the lone pair on the C, O or S, respectively, can be donated to the pi system. These two electrons from the lone pair, plus the four electrons from the two pi bonds equals six total pi system electrons. This is a number of electrons that will work for aromaticity on the basis of the Hückel rule:

$$6 \text{ electrons} = 4(1) + 2$$
$$(n = 1)$$

The answer is D.

21. *Necessary Knowledge and Reasoning:*

a. The starting material is an acid anhydride, a very reactive substrate for nucleophilic addition.

b. The reactant, $HN(C_3H_8)_2$, is moderately good nucleophile.

c. Once the nucleophile adds to the carbonyl, we have the tetrahedral intermediate with a negative charge on the formerly carbonyl oxygen:

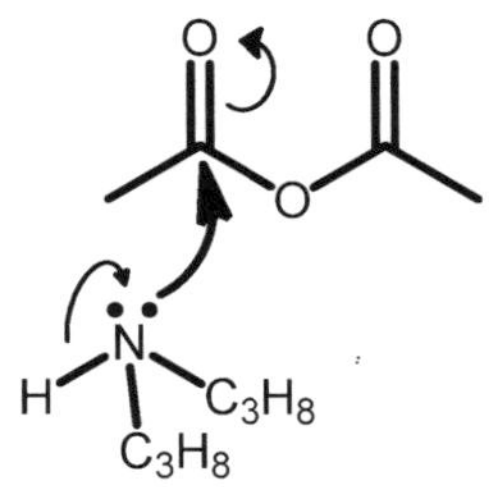

d. Recall that if there is a good leaving group on the carbon to which an O is attached, the species is subject to nucleophilic elimination to reform the carbonyl C=O bond.

e. In this case, the acetate group ($^{-}$O-C(O)CH_3) is a good leaving group because it is an anion with the negative charge on an electronegative element (O), and it is also stabilized by resonance. This means that nucleophilic elimination will occur as follows:

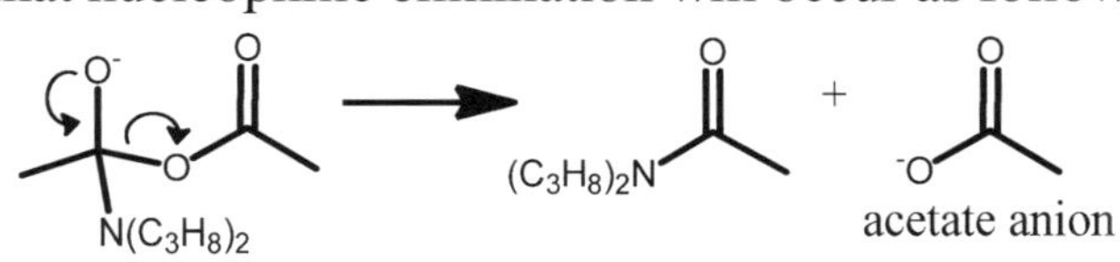

This produces the product given in answer choice C.

The answer is C.

22. *Necessary Knowledge and Reasoning:*

a. The starting materials are a carboxylic acid and an alcohol in the presence of an acid.

b. These are the conditions required for an acid-catalyzed esterification.

c. In this type of esterification, the –OH of the carboxylic acid is replaced by the –OR of the alcohol. In this case, we will get the ester in which the $–OC_2H_5$ group (–OEt group) replaces the –OH. This is the product shown in choice A.

The answer is A.

23. *Necessary Knowledge and Reasoning:*

a. The starting materials are an ester and an amine.

b. An amine is a weak base.

c. These conditions and reagents are a good combination for a base-catalyzed formation of an amide from an ester.

d. In this type of reaction the –OR of the ester is replaced with the $–NR_2$ group of the amine. Here, this means that the $–OCH_3$ will be replaced by a $–N(H)C_3H_8$ group. This matches the product in answer choice C.

The answer is C.

24. *Necessary Knowledge and Reasoning:*

a. The ketone starting material can react with nucleophiles, but there are no good leaving groups on the carbonyl carbon of the ketone.

b. Under acidic conditions, a ketone will react with two alcohols to form a ketal:

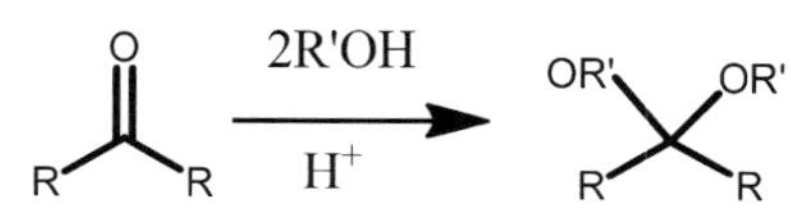

c. In the current reaction, the two alcohol groups are on the same molecule. This leads to the two alcohol O atoms attaching to the formerly-carbonyl carbon, but the two O atoms from the alcohols are of course still linked by the chain that held them together in the starting material:

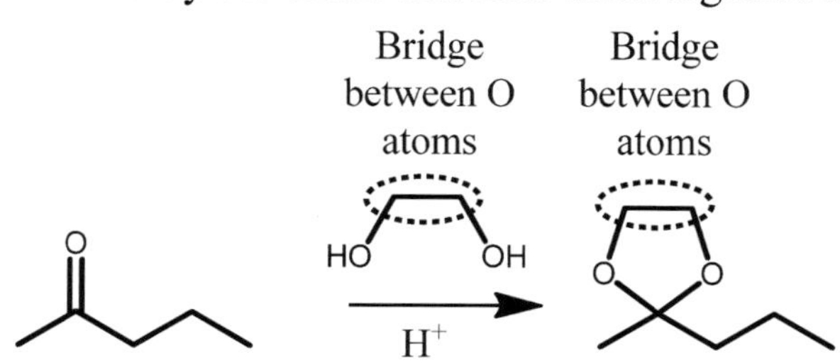

This matches the product shown in choice B.

The answer is B.

32. *Necessary Knowledge and Reasoning:*

a. The Grubbs catalyst is a ruthenium-containing catalyst that is capable of alkene metathesis (sometimes called olefin metathesis because alkenes used to be more commonly called olefins).

b. ROMP stands for **R**ing-**O**pening **M**etathesis **P**olymerization. This is an alkene metathesis reaction where the starting material is an alkene that is in a ring and the product is a polymer.

c. Alkene metathesis is the rearrangement of C=C double bonds to make new C=C double bonds by "swapping" double bonds between two alkenes:

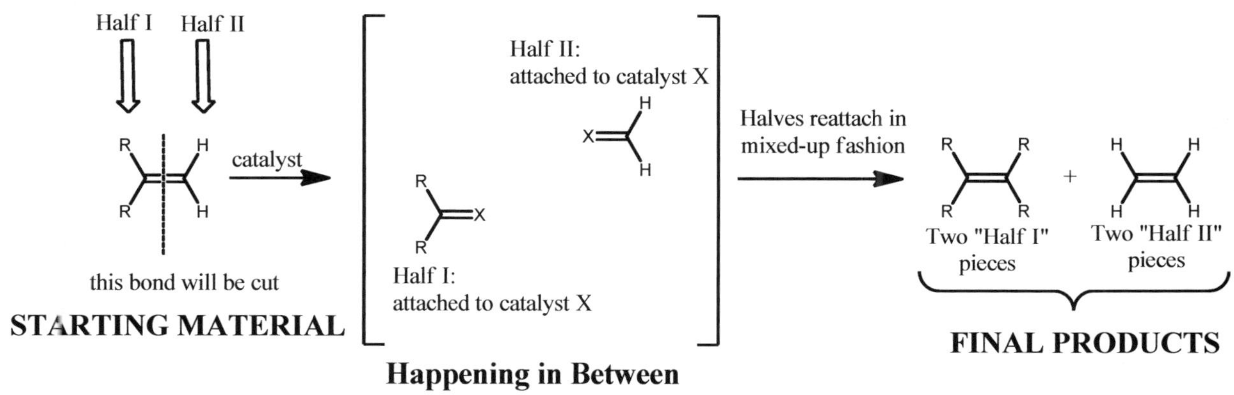

d. When the alkene to be metathesized is in a ring (like in ROMP), the reaction can be driven by relief of ring strain. The ring will thus open when the C=C bond is cut during metathesis, and it will reconnect to make a C=C bond with a carbon from another opened ring:

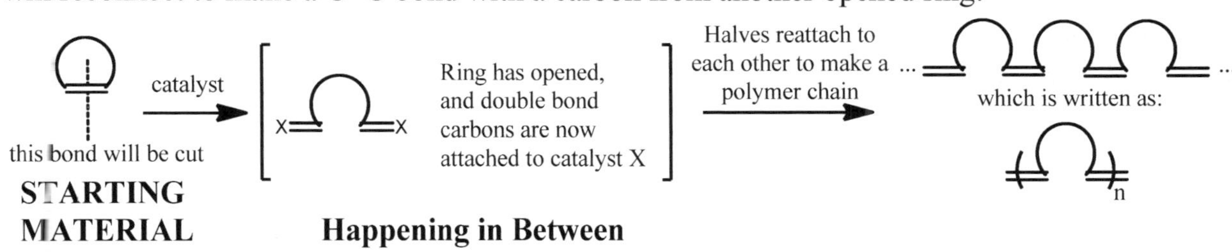

The little arc-shaped piece in this figure represents a ring that could be several different sizes.

e. In this question, we are given the product and asked to determine what the starting material was. Let us try to do this by comparing the generic scheme above to the specific case of this problem:

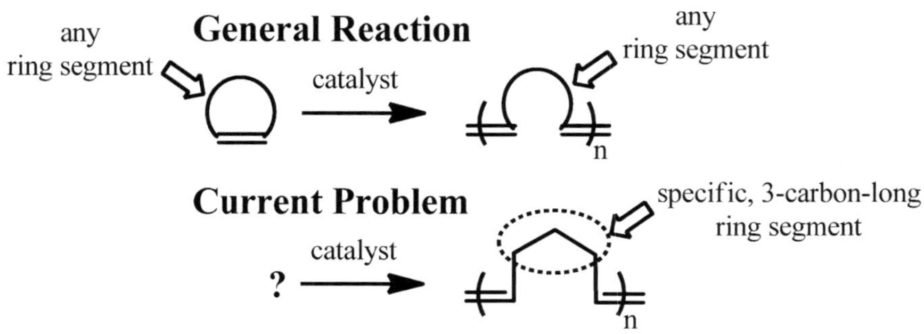

(Mixture of *cis*- and *trans*-)

e. Once we extrapolate the very general scheme to a specific example, we can see that what changes is the length of the ring segment joining the two carbons in the C=C bond. With a 3-carbon-long linker like we have in this product, we must start with cyclopentene:

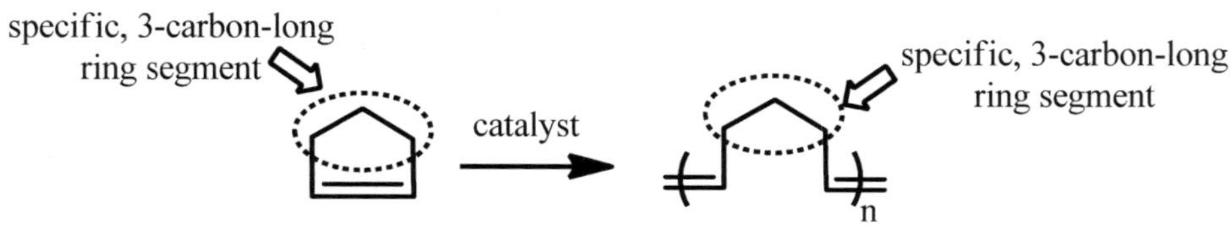

The answer is B.

26. *Necessary Knowledge and Reasoning:*

a. The reactant, P_2O_5, is a powerful desiccant, meaning that it removes water (H_2O) from a compound.

b. In the case of an amide, the H atoms and one O atom needed to make the water to be removed by the P_2O_5 comes from the carbonyl oxygen and the two H atoms on the N. This leaves the C with two few bonds and the N with two few bonds. The C and N therefore form two new pi bonds between one another to make up for the lost atoms, which results in:

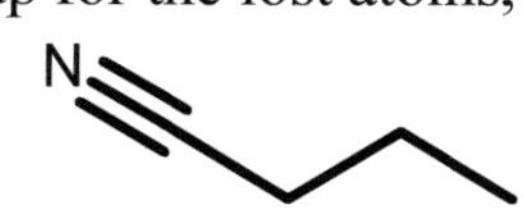

This matches the product shown in choice D. Note that this is a description of the net result of the reaction. The mechanism is several steps and should be reviewed in a reaction mechanisms book.

The answer is D.

27. *Necessary Knowledge and Reasoning:*

a. The structure of benzonitrile is:

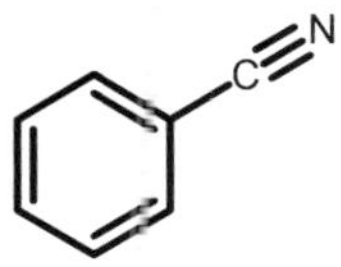

b. Acid-catalyzed hydrolysis of a nitrile group will produce a carboxylic acid, wherein the carbon that was triply-bonded to the nitrile nitrogen will be the carboxylic acid carbon.

c. We need to know the structures of the answer choices to select the correct answer. The structures are:

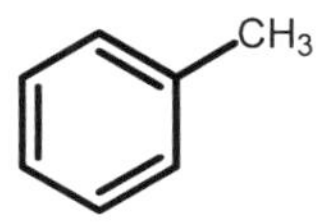

Benzaldehyde

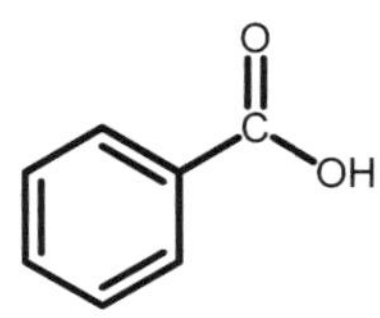

Benzoic acid

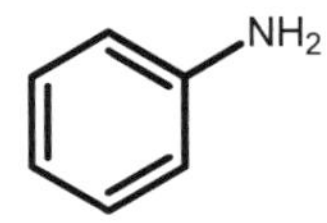

Aniline

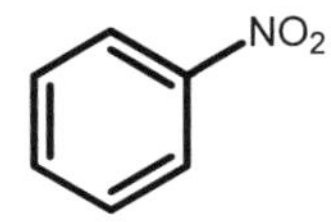

Nitrobenzene

Benzoic acid (choice B) is the only carboxylic acid.

The answer is B.

28. *Necessary Knowledge and Reasoning:*

a. An aromatic ring with BOTH an electron withdrawing group and a good leaving group that is either *ortho* or *para* to the withdrawing group is the set of conditions necessary for a nucleophilic aromatic substitution reaction.

b. A nitro- group ($–NO_2$) is a strong resonance withdrawing group. None of the other substituents on any of the choices are strong withdrawing groups.

c. Bromide, chloride and iodide are all acceptable leaving groups for this reaction. This is because these halides are all large anions and therefore very stable.

d. The only choice in which the criteria for nucleophilic aromatic substitution are met is choice B.

The answer is B.

29. *Necessary Knowledge and Reasoning:*

a. In a Suzuki reaction, a boron reagent, R-B(OR')$_2$, reacts with a reagent that has a good leaving group on it, R''-LG (R'' = vinyl or aryl), to form R–R''.

b. In this reaction, the R and R'' groups will not rearrange.

c. The piece with which we are provided in the reaction has a leaving group on it, so the missing piece we seek has a boron-containing unit in it. This rules out choices C and D.

d. In the product, the aromatic group is *meta-* to the ketone unit. We need a starting material in which the boron-containing group is *meta-* to the ketone. These criteria are met by choice B.

The answer is B.

30. *Necessary Knowledge and Reasoning:*

a. The best way to quickly assess what reaction is happening here would be to recognize the very specific and unusual reagent, phthalimide:

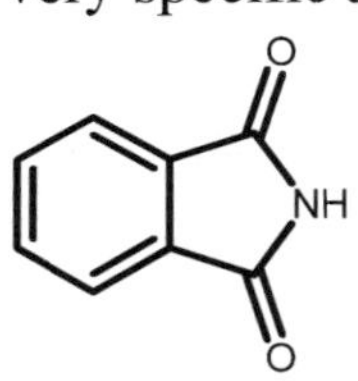

PHTHALIMIDE: USED FOR THE GABRIEL SYNTHESIS

b. Phthalimide is the starting material for a Gabriel Synthesis, which is used to make primary amines (R-NH$_2$). The R group comes from the alkyl halide, here $CH_3CH_2CH_2I$.

c. This would lead us to quickly determine that the product would be $CH_3CH_2CH_2NH_2$, 1-propyl amine (choice D).

d. We do not necessarily need to immediately recognize that this is a Gabriel synthesis, however, to work our way to the correct answer. If the starting material reacts with a base, we would expect the most acidic proton on the molecule to be removed.

e. The most acidic proton is the one on N. This is because all the atoms with protons on them in this molecule are in the same row and N is the most electronegative. The anion that will be produced by deprotonation of the N is also stabilized by resonance:

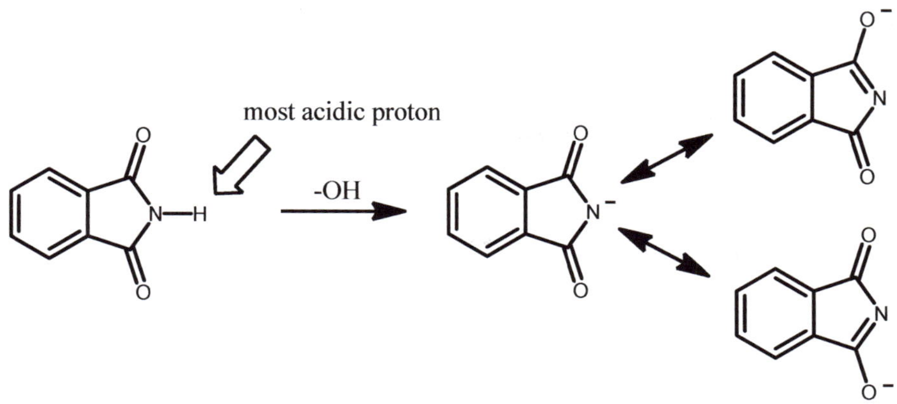

f. Once we have this anion, the next step is to react it with the 1-iodopropane. When an anion reacts with a molecule that has a good leaving group on an sp^3-hybridized carbon, we would expect an S_N2 reaction to take place:

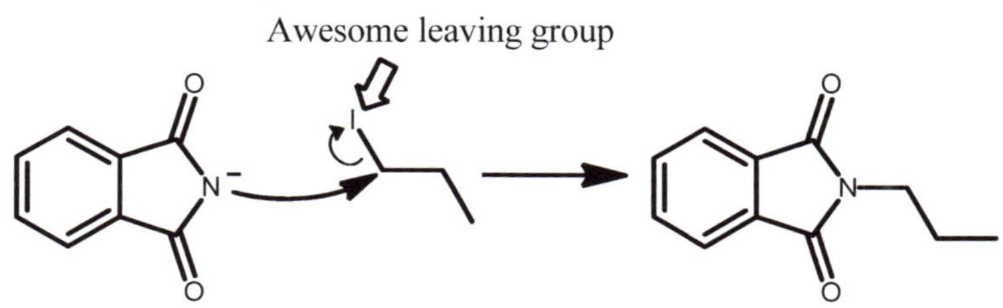

g. The next step is adding aqueous acid (HCl). Under these conditions, amide bonds (bond from N to a carbonyl carbon) will be cleaved to make an amine and a carboxylic acid:

General Reaction:

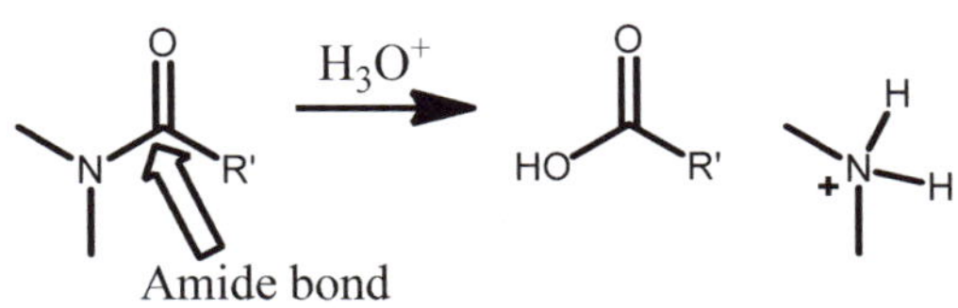

In this problem:

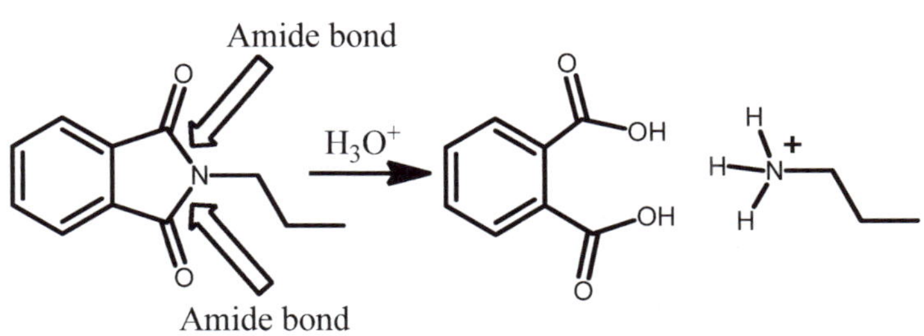

Notice a couple things here:

First, the amine product is protonated so that there is a positive charge on the nitrogen. A species with a positive charge on nitrogen is called an **ammonium** salt. Protonation occurs because the solution is acidic and an amine is a weak base.

Second, there are TWO amide type bonds in our example, so we have to break both of them.

h. The final step is to add a base to remove the proton from the ammonium, creating a neutral amine with a propyl group on it (1-propyl amine).

The answer is D.

Progress Check 3:

Material Coverage:

"Organic Chemistry 2 Lecture Guide 2013":

Lecture Sets 1 and 4-10

Date by which material should be studied:

Chapter(s) in text covering material:

Date this problem set is due:

Points this problem set is worth:

PROGRESS CHECK 3

1. Which has the highest boiling point?
(A) pentane
(B) 1-butanol
(C) 3-methylpentane
(D) propyl methyl ether

2. Which is the best nucleophile?

(A) methylamine
(B) methanol
(C) fluoromethanol
(D) fluoromethylamine

3. Which is the ***strongest*** acid?

(A) 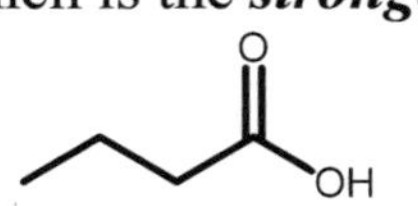 **(B)** 

(C)  **(D)** 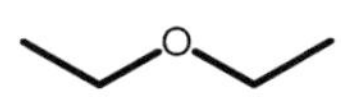

4. Which is a chiral molecule?

(A) 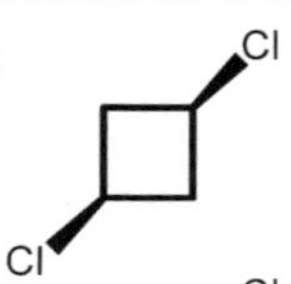 **(B)** 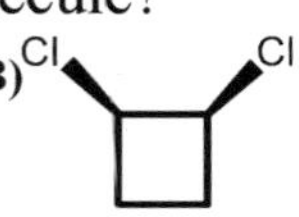

(C) 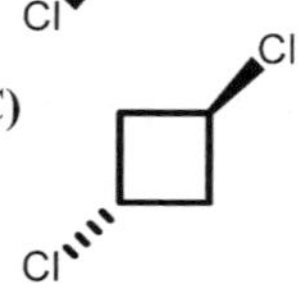 **(D)** 

5. Which is a meso compound?

(A) 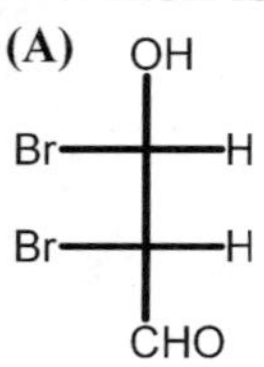 **(B)** 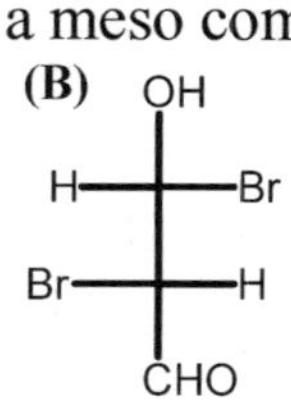 **(C)** 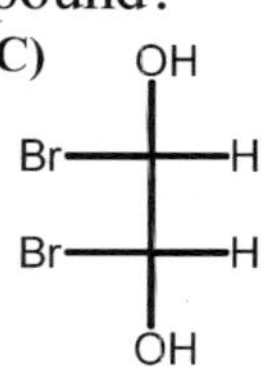 **(D)** 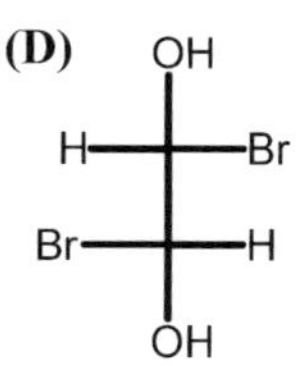

6. Which of these would be least likely to act as a dienophile in a Diels-Alder reaction?

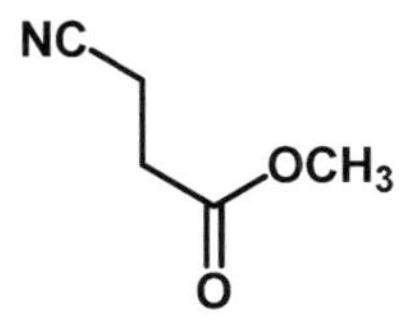
(A)

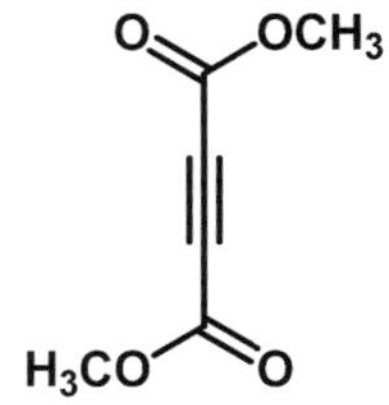
(B)

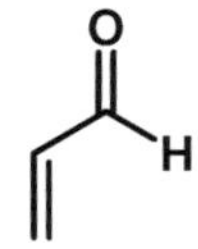
(C)

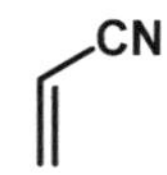
(D)

7. Which reaction should one use to make a Markovnikov alcohol from an alkene if one wants to avoid carbocation rearrangement?

A) Oxymercuration/Reduction
B) Hydroboration/Oxidation
C) Hydration
D) None of these makes an alcohol from an alkene

8. What is the product of this reaction?

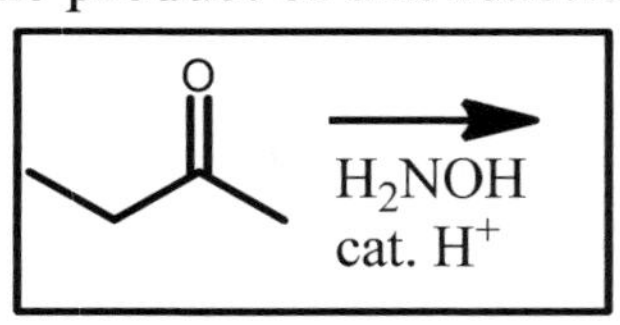

(A) 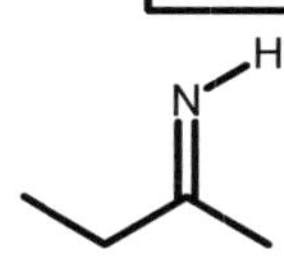 **(B)** 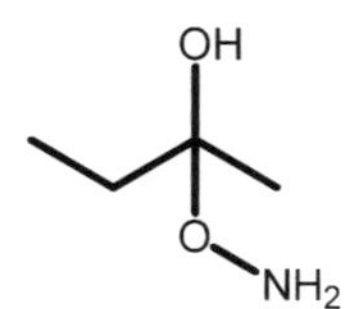

(C) 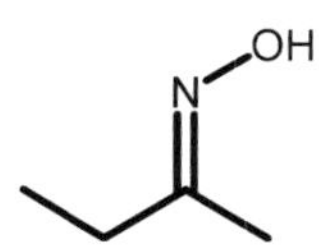 **(D)** 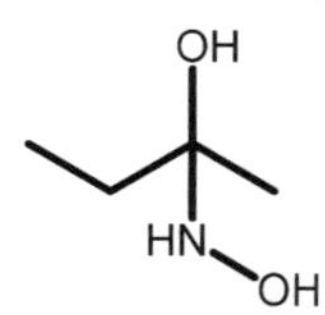

9. What is the major product of the reaction sequence?

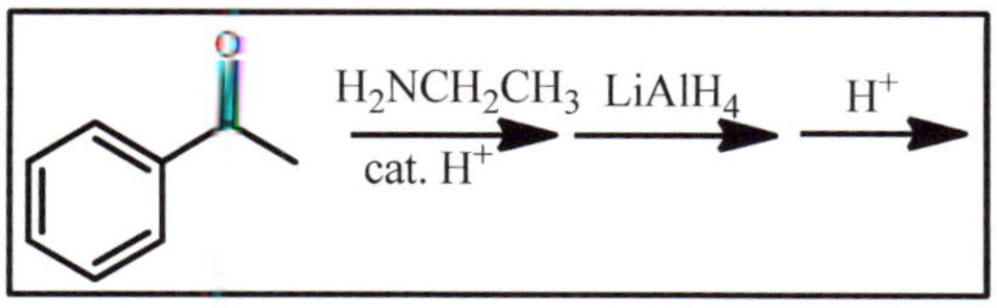

(A) 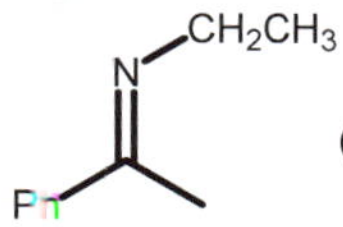

(B) 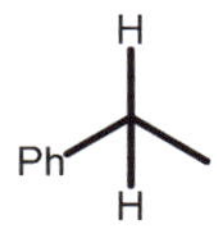

(C) 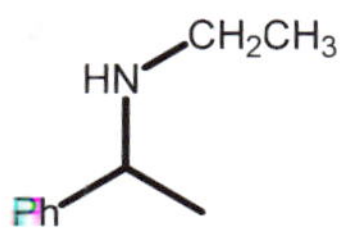

(D) 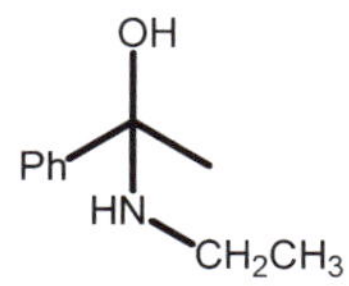

10. What is the major product of the reaction sequence?

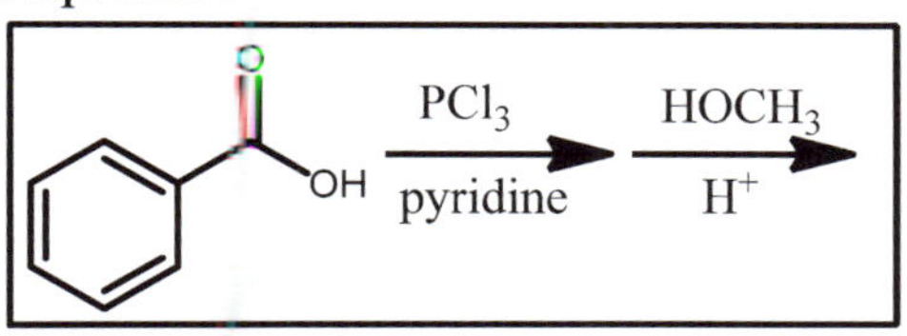

(A) 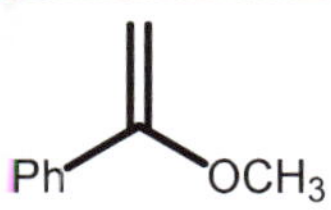

(B) 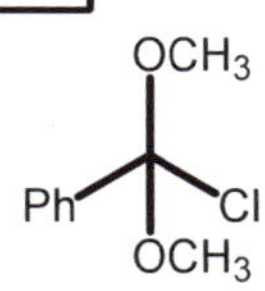

(C) 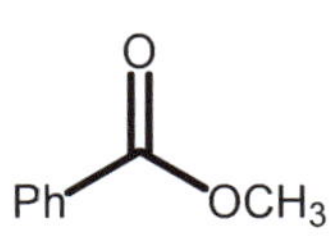

(D) 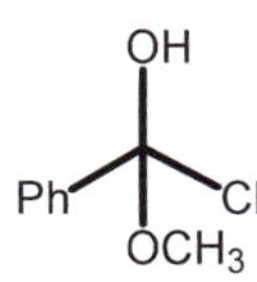

11. What is the major product of the reaction sequence?

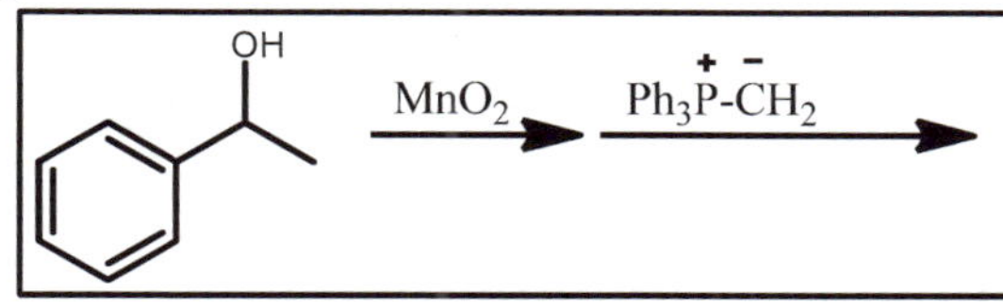

(A) 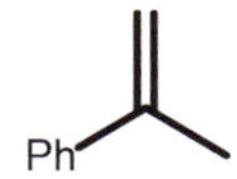

(B) 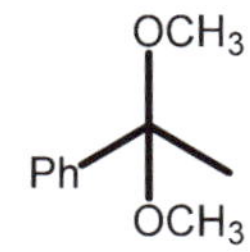

(C) 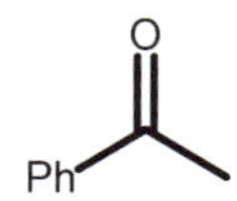

(D) 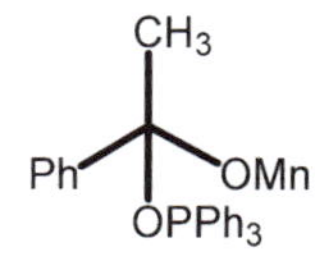

12. Which reagent is most likely to react with an α,β-unsaturated ketone to give direct addition?

A) nBuLi **B)** NaCN
C) NaSH **D)** NaN_3

13. What functional group(s) is/are found in the product when two acetone molecules undergo an Aldol **addition**?

A) a ketone without an alkene or alcohol unit
B) a ketone with an alcohol unit
C) an α,β-unsaturated carbonyl (ketone with an alkene)
D) an alcohol without an alkene or ketone unit

14. What functional group(s) is/are found in the product when two acetone molecules undergo an Aldol **condensation**?

A) a ketone without an alkene or alcohol
B) a ketone with an alcohol unit
C) an α,β-unsaturated carbonyl (ketone with an alkene)
D) an alcohol without an alkene or ketone

15. What is the major product of this reaction?

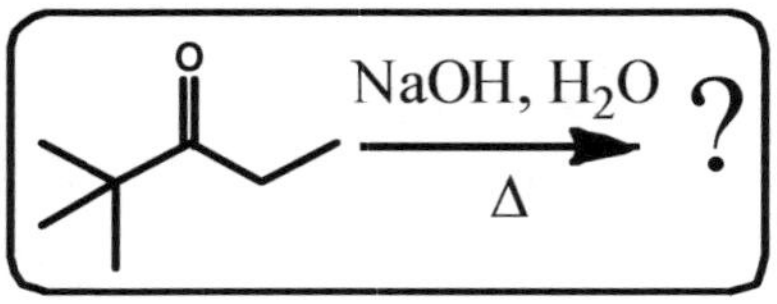

(A) 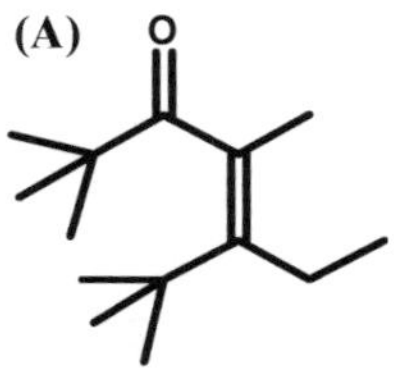

(B) 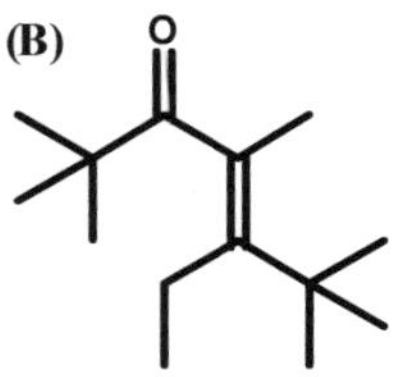

(C) 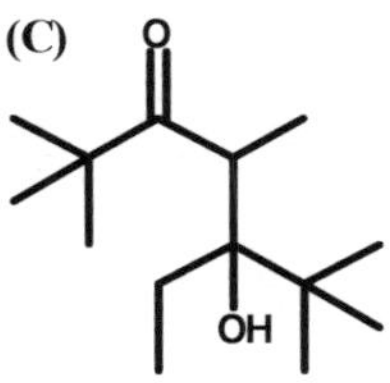

(D) 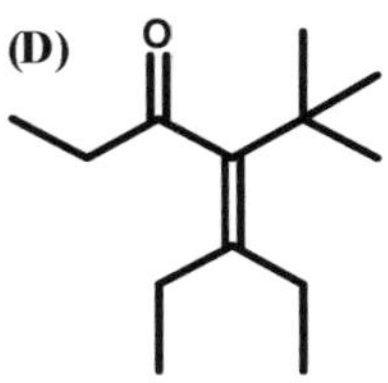

16. What is the major product of this Michael reaction?

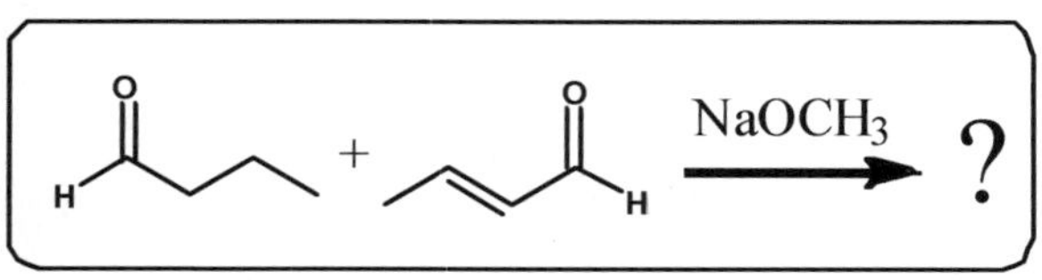

(A) 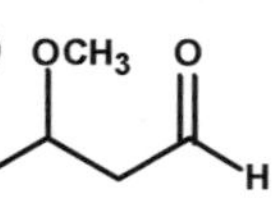

(B) 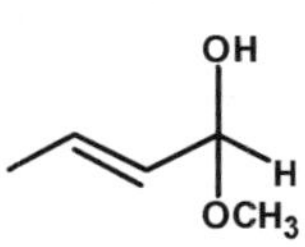

(C) 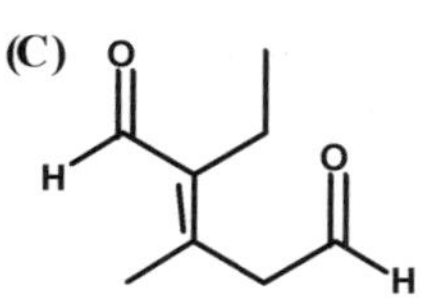

(D) 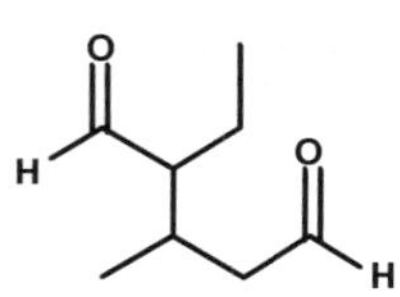

17. Which R-Br reacts fastest in this S_N2 reaction?

(A) 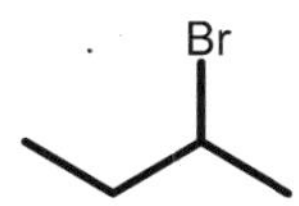

(B) 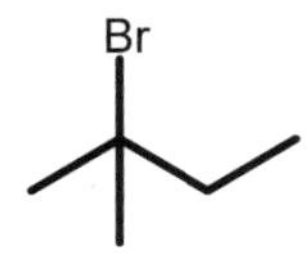

(C) 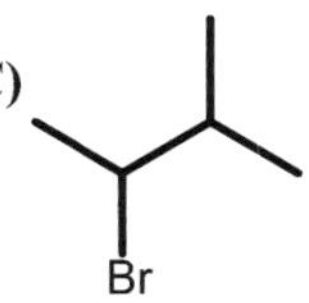

(D) 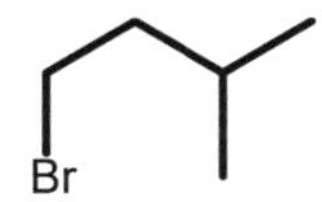

18. The major substitution product(s) of the following reaction will be:

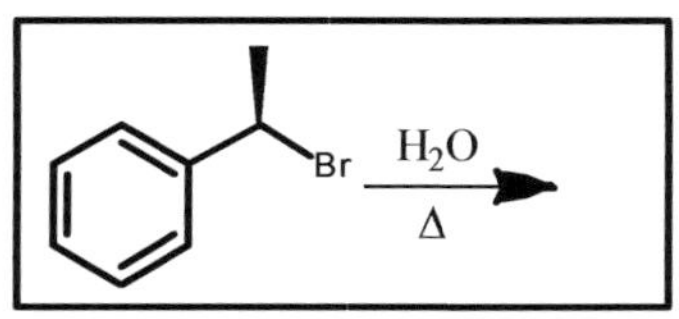

(A) predominantly *R*-1-phenylethanol
(B) predominantly *S*-1-phenylethanol
(C) A racemic mixture of *R*- and *S*-1-phenylethanol
(D) No reaction

19. The major substitution product(s) of the following reaction will be:

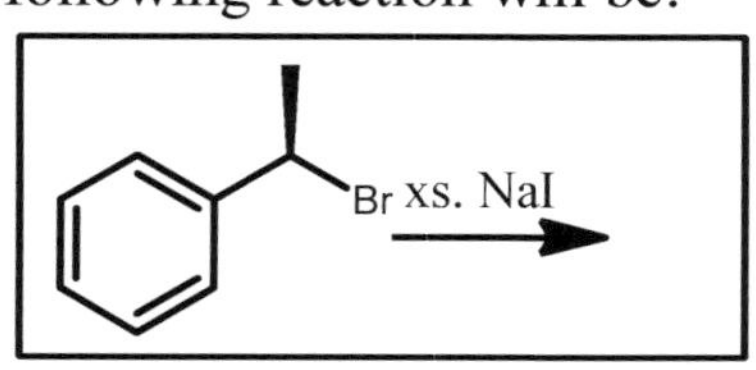

(A) predominantly *R*-1-phenylethanol
(B) predominantly *S*-1-phenylethanol
(C) A racemic mixture of *R*- and *S*-1-phenylethanol

(D) No reaction

20. What is a major product of this reaction?

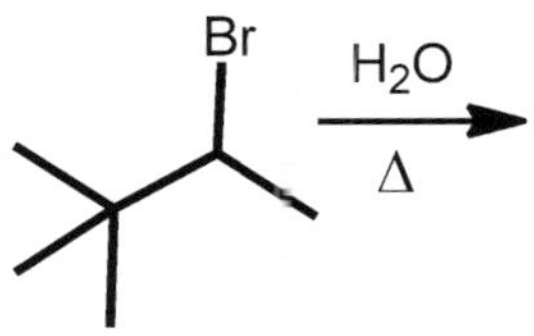

(A)

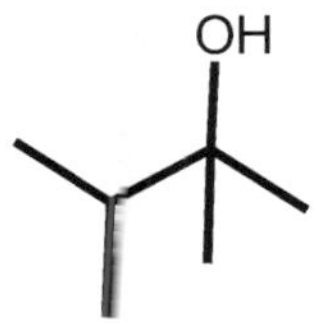

(B)

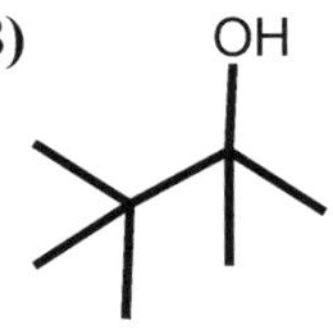

(C)

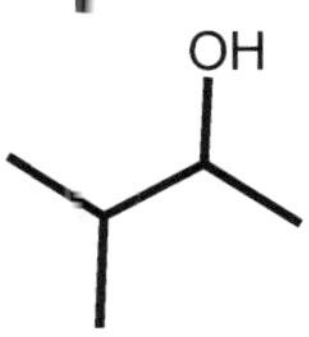

(D)

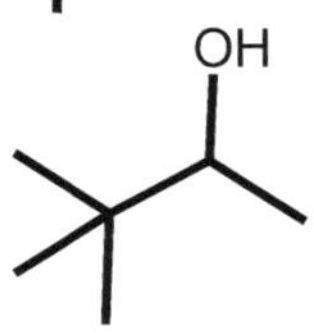

21. Provide the product of this Claisen Condensation?

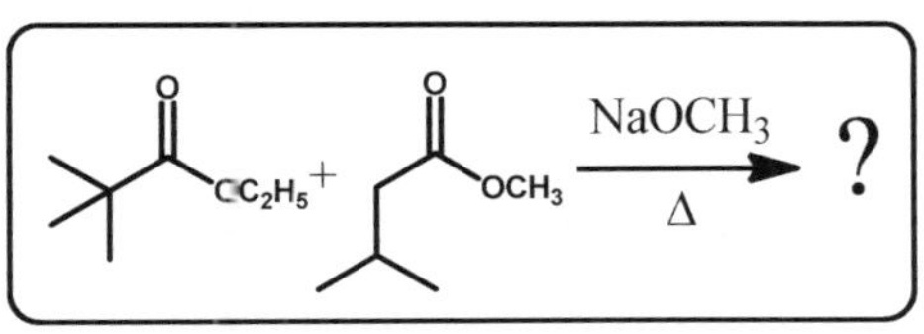

(A)

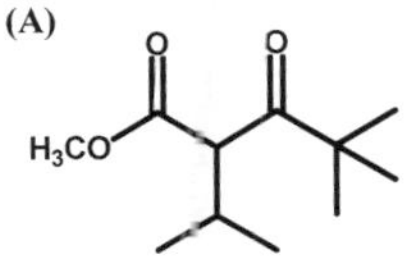

(B)

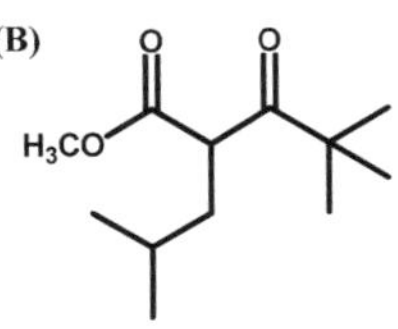

(C)

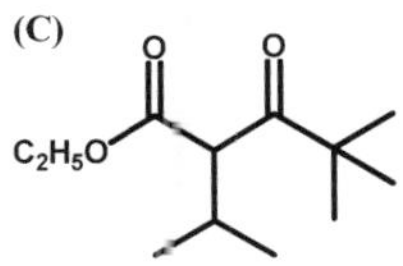

(D)

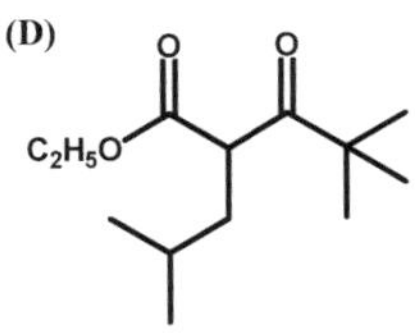

22. What is the major product of this reaction?

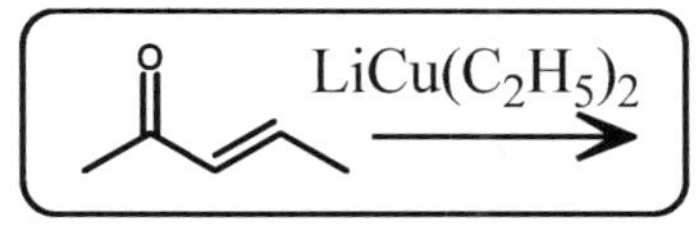

(A)

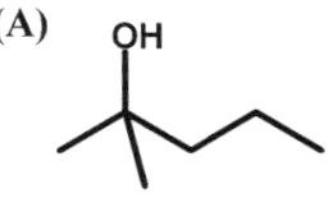

(B)

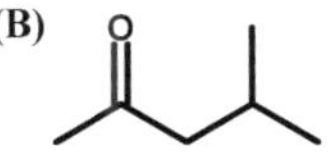

(C)

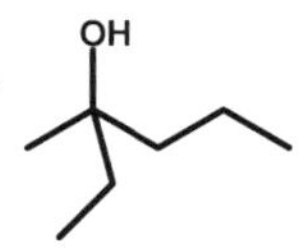

(D)

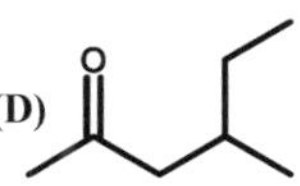

23. Which of these could be correctly named as an *E*- alkene?

(A)

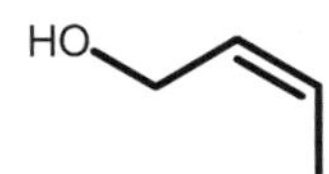

(B)

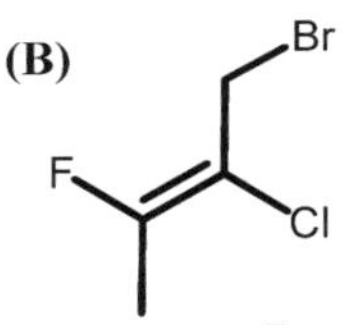

(C)

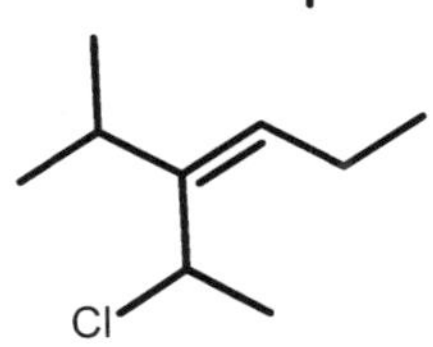

(D)

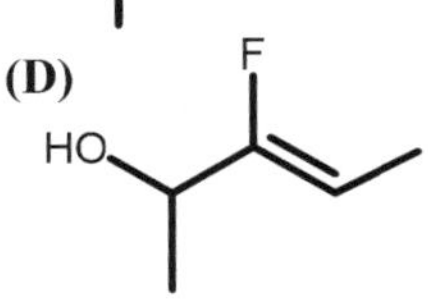

24. Which proton in this compound is attached to a *sp*-hybridized carbon?

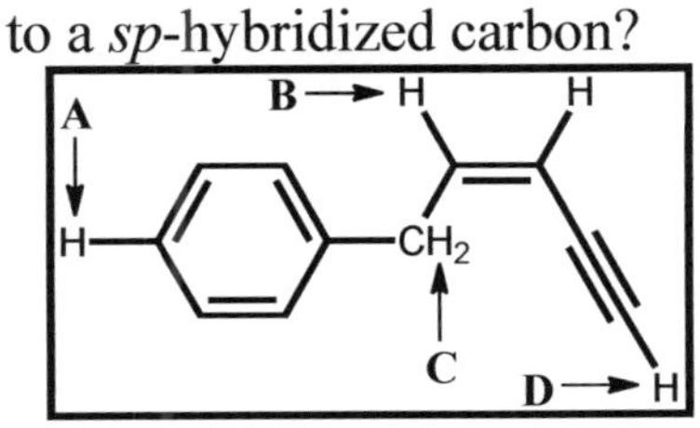

(A) A
(B) B
(C) C
(D) D

25. What conditions will give the best yield of an alkene from an alkyne?

(A) Grignard reagent then acid
(B) Lindlar's catalyst and H_2
(C) H_2/Pt or H_2/Pd
(D) H_2O/H^+

26. What diene is needed to prepare the product shown via a Diels-Alder reaction?

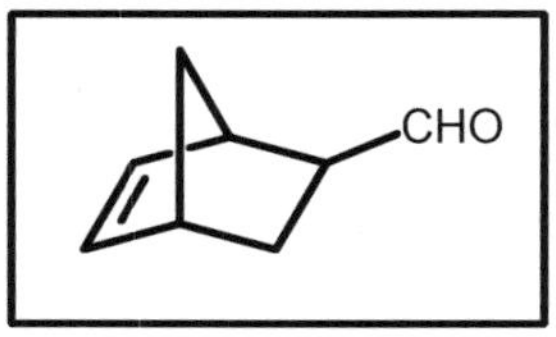

(A) 

(B) 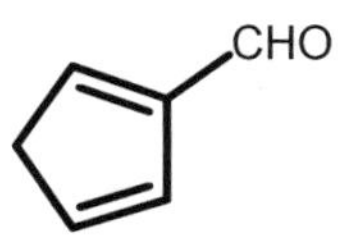

(C) 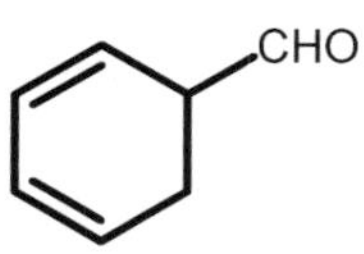

(D) 

27. What is the relationship between these two structures?

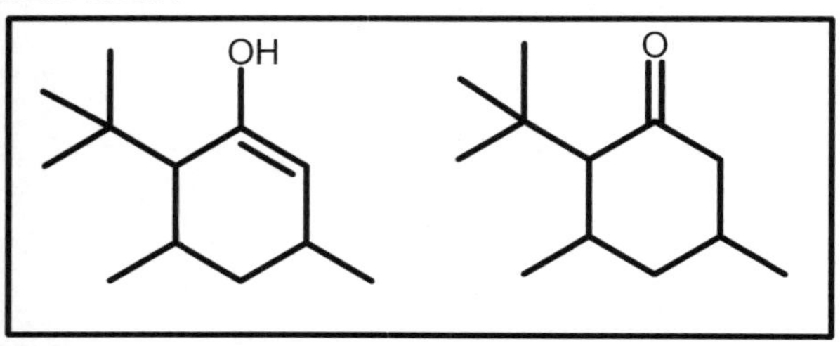

(A) Enantiomers
(B) Resonance Structures
(C) Configurational isomers
(D) Tautomers

28. What is the major product of the reaction sequence?

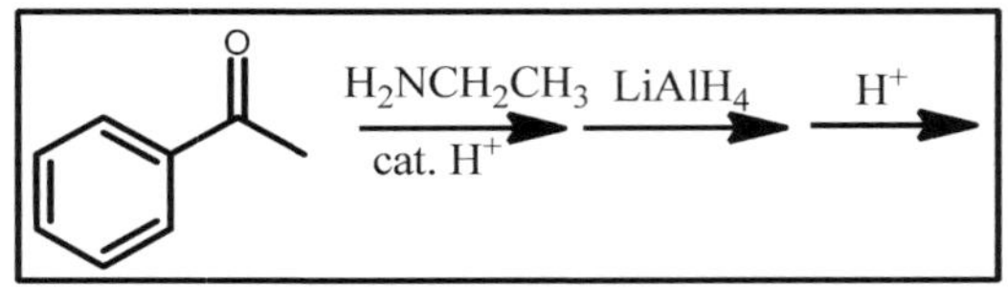

(A) 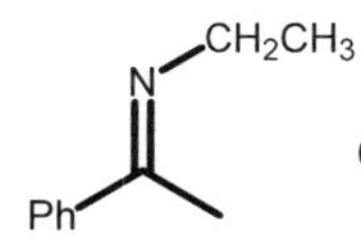

(B) 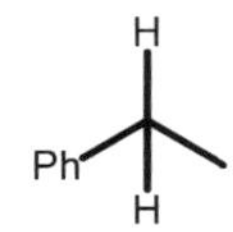

(C) 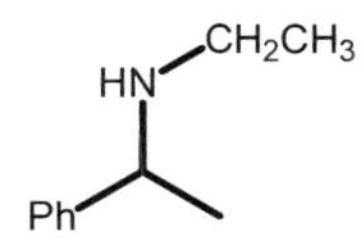

(D) 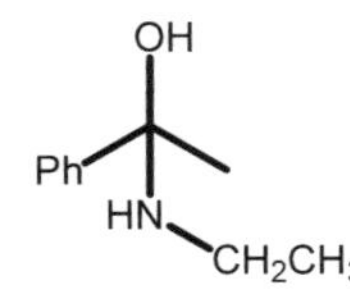

29. Which is the most basic compound?

(A) 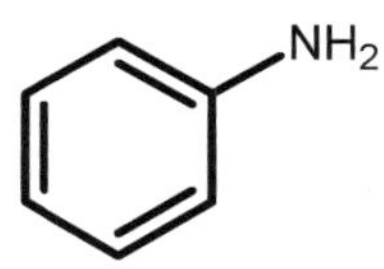

(B) 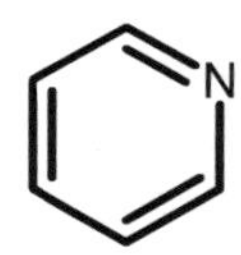

(C) 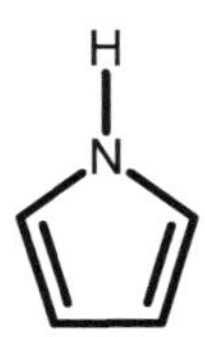

(D) 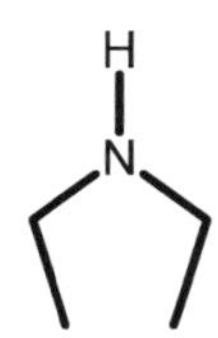

30. What is the major product of the following reaction?

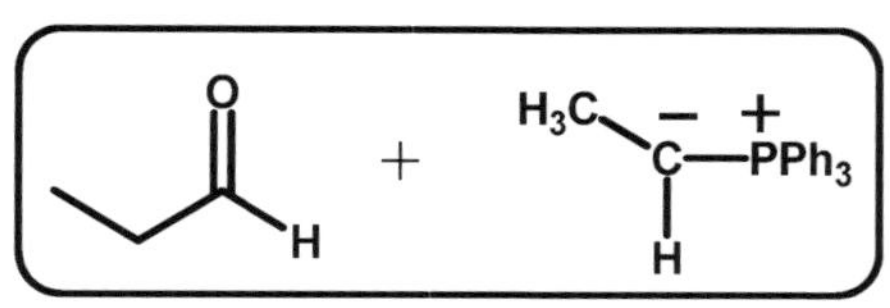

(A) (*Z*)-2-pentene
(B) (*E*)-2-pentene
(C) Propanol

(D) Pentanol

31. Which is the major product of the indicated reaction sequence?

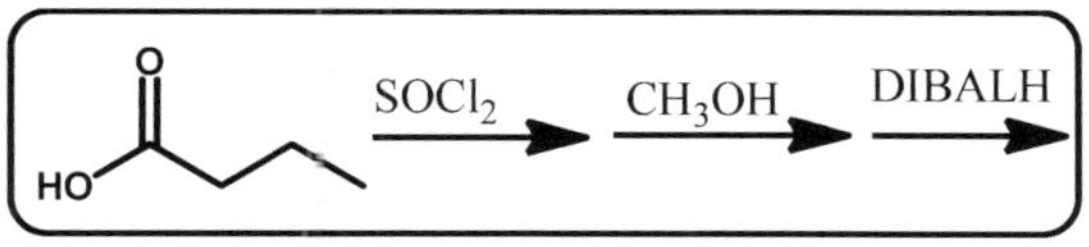

(A)

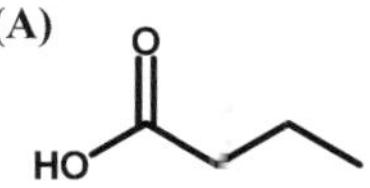

(B)

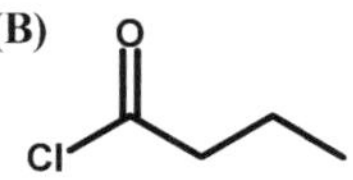

(C)

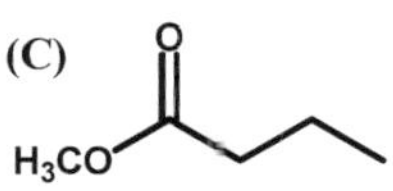

(D)

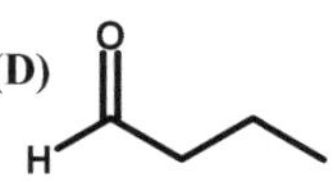

32. Which of the following can undergo decarboxylation upon heating in the presence of acid?

(A)

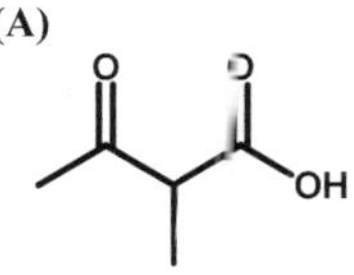

(B)

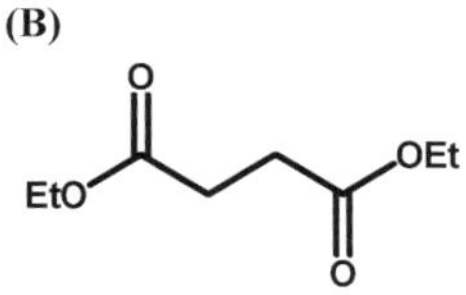

(C)

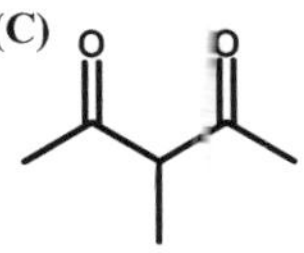

(D)

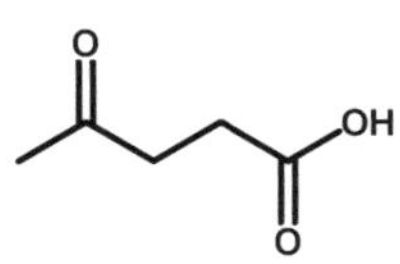

33. Deuterium (D) is an isotope of hydrogen. Any time deuterium appears in a reactant, it will end up in the product in the same position as a hydrogen (H) would appear. On the basis of this information, provide the product of the following reaction:

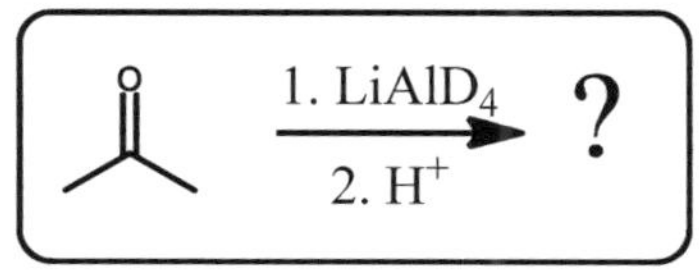

(A)

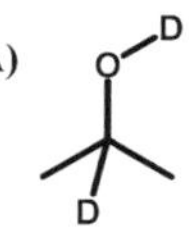

(B)

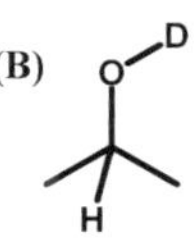

(C)

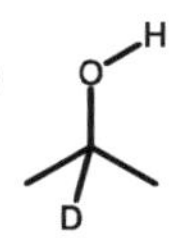

(D)

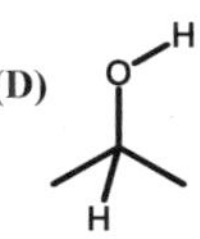

34. Provide the product of this diazotization reaction:

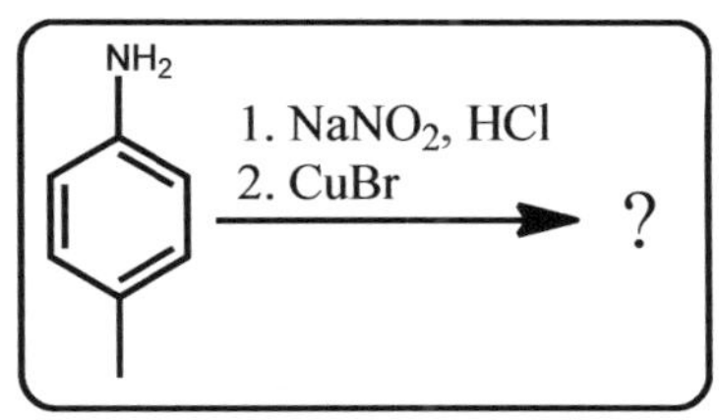

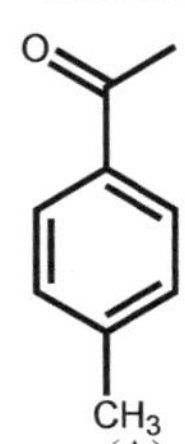

(A)

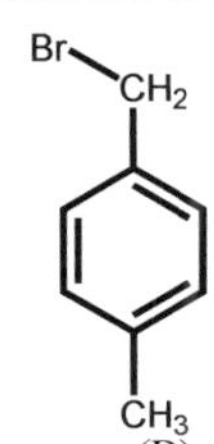

(B)

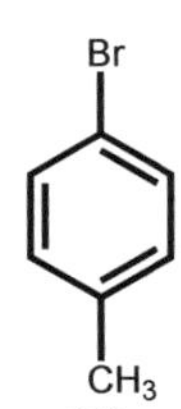

(C)

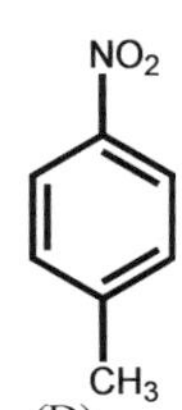

(D)

35. Which alkyl halide is required as a step in the malonic ester synthesis to yield the major product shown?

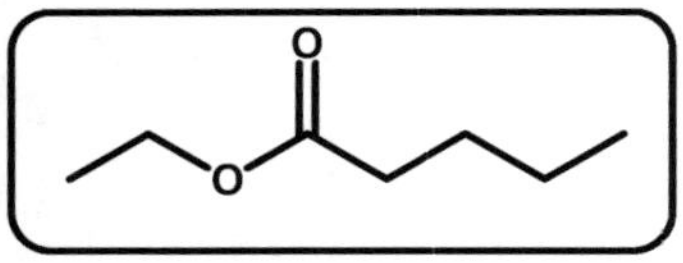

A) iodomethane
B) iodoethane
C) 1-iodopropane
D) 1-iodobutane

Key for Progress Check 3:

1. B
2. A
3. A
4. D
5. C
6. A
7. A
8. C
9. C
10. C
11. A
12. A
13. B
14. C
15. B
16. D
17. D
18. C
19. B
20. A
21. A
22. D
23. B
24. D
25. B
26. D
27. D
28. C
29. D
30. B
31. D
32. A
33. C
34. C
35. C

Detailed Feedback for Progress Check 3

For each question that you missed or were unsure of, review the Necessary Knowledge and Reasoning and possible mistakes provided below. By addressing the gaps in your knowledge and avoiding similar mistakes on future problems, you should see improvement in your scores.

1. *Necessary Knowledge and Reasoning:*

a. The boiling point is the temperature at which a liquid becomes a gas

b. Molecules in the gaseous phase are farther apart than are molecules in the liquid phase. In fact, the intermolecular forces between gas-phase molecules are negligible because the molecules are so far apart.

c. The stronger the intermolecular forces between molecules in a sample of a liquid, the more thermal energy (higher temperature) will be needed to break them apart into a gaseous state.

d. Hydrogen bonding is the strongest intermolecular force present between neutral organic molecules. Hydrogen bonding is possible in molecules that have H-O, H-N, or H-F bonds. Choice B is the only one that is capable of hydrogen bonding.

The answer is B.

2. *Necessary Knowledge and Reasoning:*

a. A nucleophile is something that donates electrons to an electron acceptor (like an electrophile).

b. The more electronegative an element, the more tightly it holds onto its electrons. Therefore, a less electronegative element will more easily act as a nucleophile if it has electrons that are not tied up in bonds. In this case, two choices are alcohols; an alcohol has lone pair electrons on an oxygen atom. The other two choices are amines; an amine has lone pair electrons on a nitrogen atom. Because nitrogen is less electronegative than oxygen, we expect an amine to generally be better nucleophile than an alcohol. This rules out choices B and C.

c. Next, we have to consider inductive effects. Any electronegative elements attached to the molecule *near* the nitrogen will be pulling electrons towards them (an inductive withdrawing effect). This will make it harder for the nitrogen to give the electrons to another species, so a molecule with inductively withdrawing groups on it will be a poorer nucleophile.

d. Choice D has an electron-withdrawing fluorine atom attached to the same carbon as the nitrogen, so this will be a poorer nucleophile than the other amine choice, A. Thus, choice A will be the best nucleophile.

The answer is A.

3. *Necessary Knowledge and Reasoning:*

a. The *strongest* acid is the one that produces the *most* stable anion upon removal of the proton.

b. A more electronegative element generally loses its proton to produce a more stable anion than a comparably-sized but less electronegative element. In this problem, only choices A and B have protons that you can take off of O atoms; in choices C and D, only the carbon atoms have protons on them. We can thus rule out choices C and D.

c. If the negative charge can be delocalized via resonance, it acquires greater stability as compared to an anion that lacks resonance delocalization but has the negative charge on the same atom (here, oxygen). Only choice A has an anion that can be stabilized via resonance, so that anion is most stable, making choice A the strongest acid.

The answer is A.

4. *Necessary Knowledge and Reasoning:*

a. In order for a molecule to be chiral, it must not have a plane of symmetry in it.

b. Only choice D lacks a plane of symmetry.

The answer is D.

5. *Necessary Knowledge and Reasoning:*

a. A meso compound must have stereogenic atoms in it, but must ALSO have a plane of symmetry or inversion center.

b. All of the choices have stereogenic atoms in them, so this criterion doesn't help us rule out any choices.

c. Choice C, however, is the only one with a plane of symmetry, so this is the meso compound.

The answer is C .

6. *Necessary Knowledge and Reasoning:*

a. In a Diels-Alder reaction the term dienophiles refers to something with a C-C pi bond, which reacts with a diene.

b. Choice A has no C-C pi bond in its structure and thus cannot fill the role of a dienophile.

The answer is A.

7. *Necessary Knowledge and Reasoning:*

a. Oxymercuration/reduction (choice A) leads to a Markovnikov alcohol as the major product

b. The Hydroboration/oxidation (choice B) leads to a non-Markovnikov alcohol as the major product

c. Hydration (choice C) leads to a Markovnikov alcohol, but it proceeds via a carbocation intermediate.

The answer is A.

8. *Necessary Knowledge and Reasoning:*

a. One starting material is a ketone.

b. The other reactant is a hydroxylamine (an amine with an OH on the nitrogen) with two hydrogens on the nitrogen.

c. Under acid-catalyzed conditions, ketones will often lose water, which is made from the O of the C=O bond with two protons provided by the other reactant(s). A few of the general reaction types that are usually covered in an introductory organic course are:

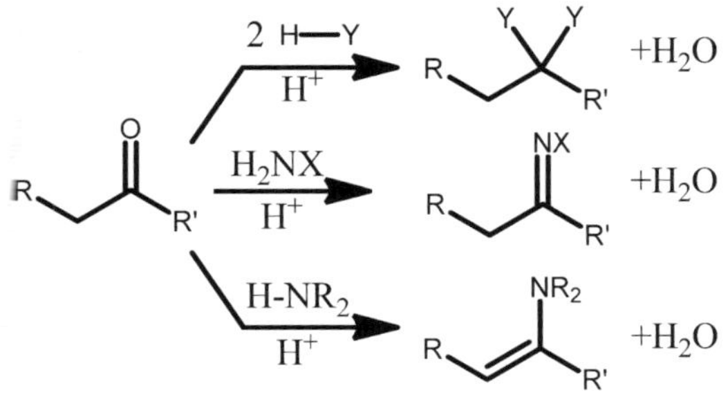

It is worth noting that aldehydes can undergo the same kind of reactions.

d. When the reactant is a ketone or an aldehyde, it is worth considering these possibilities. In the current case, our reactant features a nitrogen with TWO protons on it. This will allow for a net reaction like the second pathway shown above. We would thus expect the product to feature a double bond between nitrogen and the carbon-formerly-known-as-carbonyl:

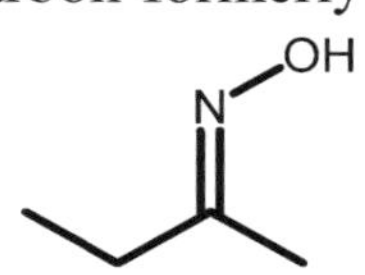

This structure matches that shown in choice C.

The answer is C.

9. *Necessary Knowledge and Reasoning:*

a. The best way to approach a multistep synthesis problem is to consider one step at a time. Here, the starting material is a ketone and the first step is to react it with a primary amine with catalytic acid (the abbreviation "cat." Indicates a catalytic amount of something).

b. Under acid-catalyzed conditions, ketones will often lose water, which is made from the O of the C=O bond with two protons provided by the other reactant(s). A few of the general reaction types that are usually covered in an introductory organic course are:

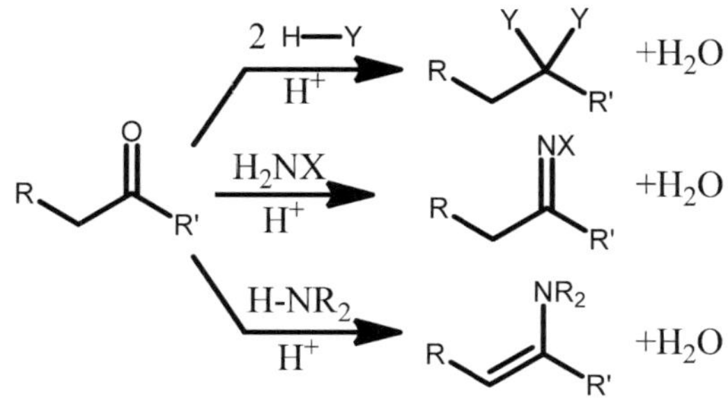

Hydrate or Ketal formation:
Example: Y = OH or OR

Double bond to Nitrogen: requires TWO protons on N in the starting material

Enamine reaction; only ONE proton on the nitrogen in the starting material

It is worth noting that aldehydes can undergo the same kind of reactions.

c. When the reactant is a ketone or an aldehyde, it is worth considering these possibilities. In the current case, our first reactant features a nitrogen with TWO protons on it. This will allow for a net reaction like the second pathway shown above. We would thus expect the product to feature a double bond between nitrogen and the formerly-carbonyl carbon:

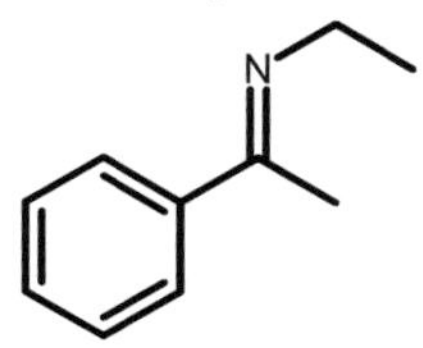

This imine (also called a Schiff base) will be the product of our first step, and will serve as the starting material for step 2.

d. Imines react very similarly to carbonyls in many cases.

e. Our second step features $LiAlH_4$ as a reactant. This species will produce hydride anions (H^-) as the putative reactive species. Hydride is a very good nucleophile, and will react with a polar pi bond by nucleophilic addition like this:

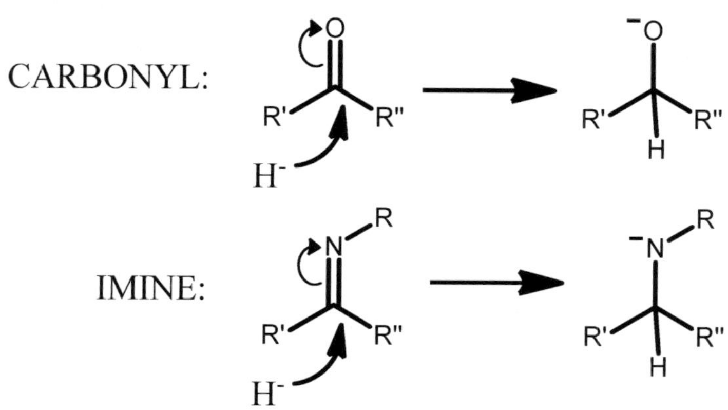

f. We would thus expect that the product of our second step would be:

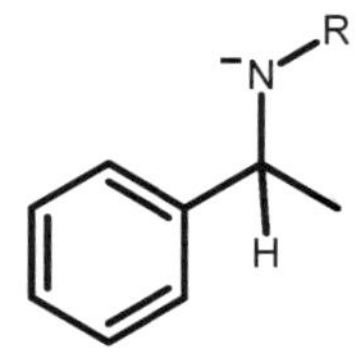

g. The final step, H^+, will simply place a proton on the anionic nitrogen to make the neutral amine, which is shown in choice C (note that Ph stands for phenyl, the name given to benzene when it is a substituent).

The answer is C.

10. *Necessary Knowledge and Reasoning:*

a. The best way to approach a multistep synthesis problem is to consider one step at a time. The first reaction is between PCl_3 and a carboxylic acid. This is a very good way to exchange the poor leaving group –OH of the carboxylic acid for a good leaving group –Cl. This is often referred to as an "activation" reaction because it turns the carboxylic acid into a much more reactive acid chloride. The product of this first step is thus:

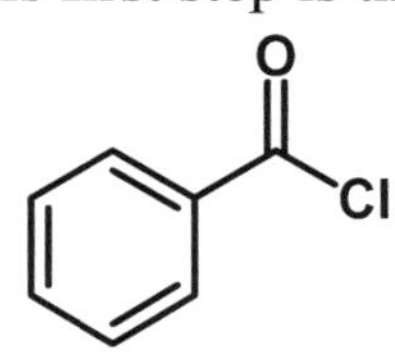

Note that this is choice B, but it is not the product of the whole series of reactions, so we do cannot choose it until we work through all the other steps.

b. Acid chlorides react readily with a wide variety of nucleophiles via a nucleophilic addition-nucleophilic elimination sequence wherein the Cl is replaced by the nucleophile by this general reaction:

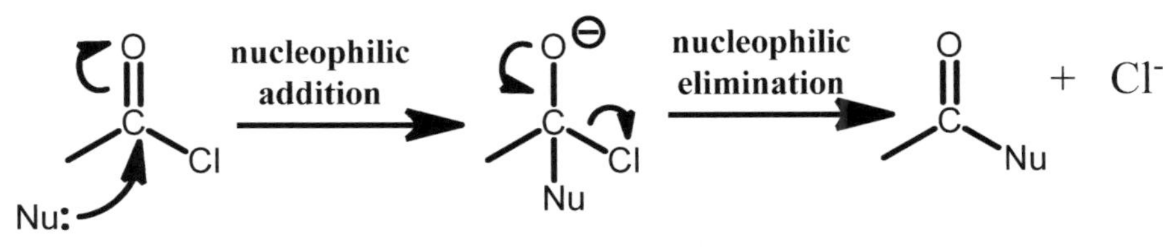

c. In this case, it is the –OCH_3 portion of the methanol reactant that will replace the Cl. This means that the product of the second step is:

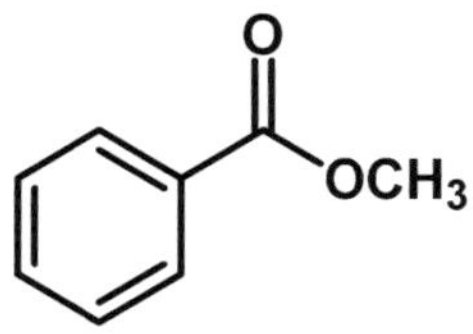

This matches the structure provided in choice C. Note that "Ph-" is an abbreviation for phenyl, the name given to benzene when it is a substituent.

The answer is C.

11. *Necessary Knowledge and Reasoning:*

a. MnO_2 is an oxidizing agent that will oxidize a primary alcohol to an aldehyde or a secondary alcohol to a ketone.

b. In this case, we have a secondary alcohol, so it is oxidized to the corresponding ketone:

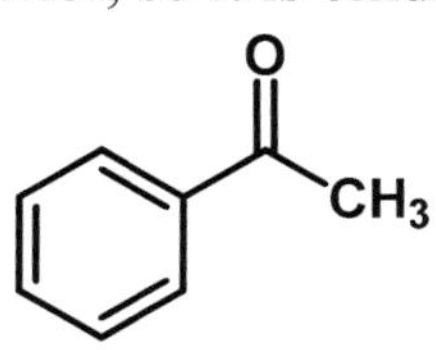

c. The second reaction features a very unusual reagent, a phosphorus ylide.

b. In introductory organic chemistry courses, the phosphorus ylide is generally only used in the Wittig reaction, which produces an alkene. In this reaction, the carbon with the formal charge of -1 will end up doubly bound to the carbonyl carbon in place of the carbonyl, and the phosphorus-containing piece ends up doubly bound to the oxygen instead:

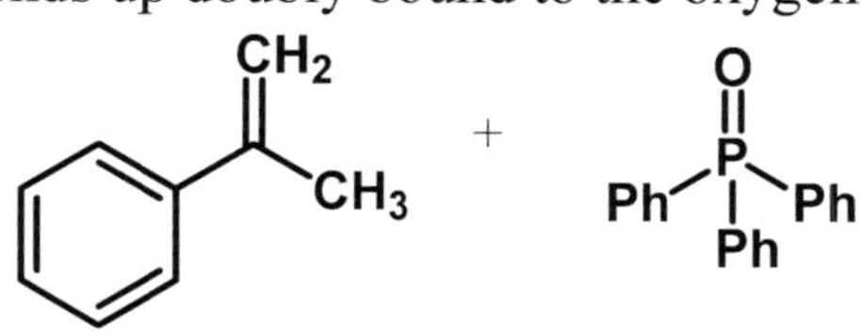

The structure on the left matches that provided in choice A..

The answer is A.

12. *Necessary Knowledge and Reasoning:*

a. Nucleophiles that are stronger bases (stronger than hydroxide) tend to undergo primarily direct addition to α,β-unsaturated ketones.

b.Nucleophiles that are weaker bases (hydroxide or weaker) tend to undergo primarily conjugate addition to α,β-unsaturated ketones.

c. The less stable the anion, the stronger base it is. We are looking for a nucleophile that will *best* undergo direct addition, so we are looking for the least stable anion.

d. The anion produced in solution for each choice will be:

Choice A: $CH_3CH_2CH_2CH_2^-$
Choice B: $^-C{\equiv}N$
Choice C: ^-SH
Choice D: $[N{=}N{=}N]^-$

e. The next step is to evaluate the different structural features of an anion that allow us to make predictions about anion stability.

e. **Size (Row) Effect:** An anion in which the negative charge is on a larger atom (a higher-numbered row of the periodic table) will be more stable than one in which the negative charge is on a smaller atom.

This effect lets us rule out choice C: ^-SH is more stable than hydroxide and is thus a weaker base.

f. **Electronegativity Effect:** An anion is more stable if it features a negative charge on a more electronegative element.

This effect lets us rule out choice D because the negative charge is on N in this choice, whereas it is on carbon in the other remaining choices.

g. **Hybridization Effect:** An anion is more stable if the negative charge is on an atom with more *s* character in its valence orbitals. This means that, all other things being equal, an anion with a negative charge on an *sp* hybridized carbon will be more stable than one with a negative charge on an sp^2 hybridized carbon, which in turn is more stable than an anion with a negative charge on an sp^3 hybridized carbon.

This rule lets us rule out choice B because the negative charge in the cyanide anion is on an *sp* hybridized carbon.

The answer is A.

13. *Necessary Knowledge and Reasoning:*

a. An Aldol **addition** results when an enolate acts as a nucleophile in nucleophilic addition to an aldehyde or ketone. For acetone, the Aldol addition reaction will look like this:

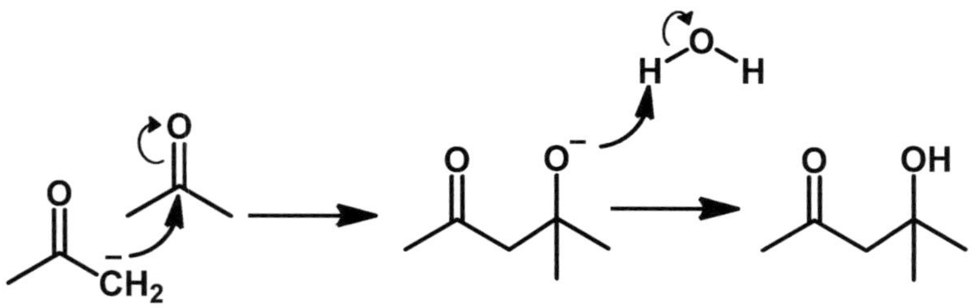

The product of this addition step features a ketone and an alcohol functional groups.

The answer is B.

14. *Necessary Knowledge and Reasoning:*

a. The Aldol **condensation** occurs when the **addition** product (shown in the previous question) is heated in the presence of a base to drive off water. This will lead to the formation of an α,β-unsaturated ketone. This is enough information to realize that the answer is C.

The answer is C.

15. *Necessary Knowledge and Reasoning:*

a. When one heats a carbonyl in the presence of a base, an Aldol Condensation can occur if the carbonyl has protons at the alpha position. In this example, the starting material has alpha protons as indicated here:

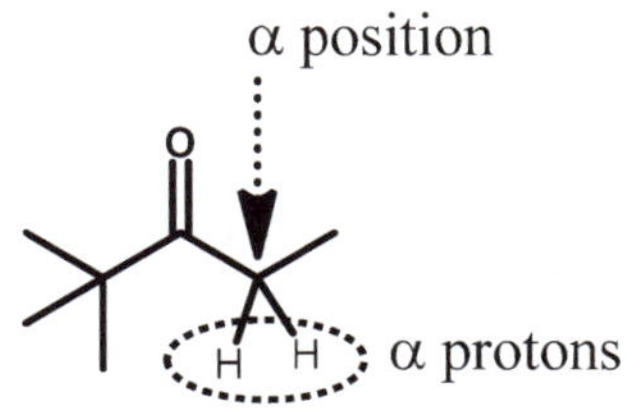

b. An Aldol condensation leads to the alpha position of one molecule ending up doubly bound to the carbonyl C in place of the double bond to O:

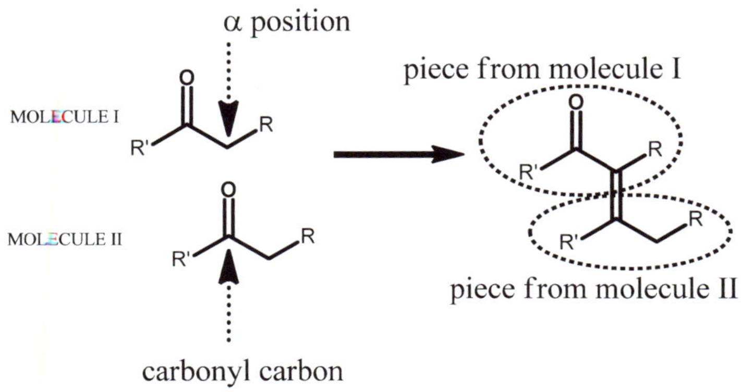

c. It is important to note that there are two ways that molecule II could attach to molecule I. In the product drawn above, the R' group of molecule II ends up on the same side as the carbonyl derived from molecule I. We could alternatively try to do the reaction so that molecule II is flipped over, like this:

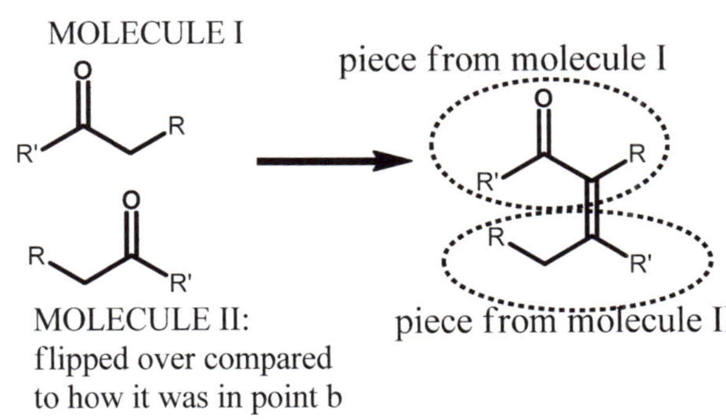

d. It is more favorable to make the less sterically hindered product.

e. In the current case, there are two possible Aldol condensation products:

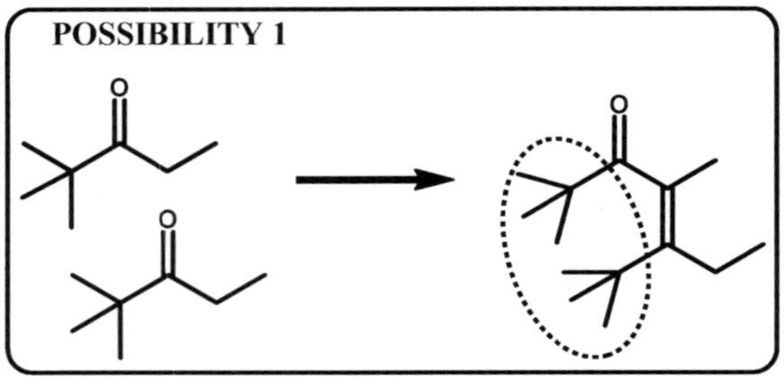

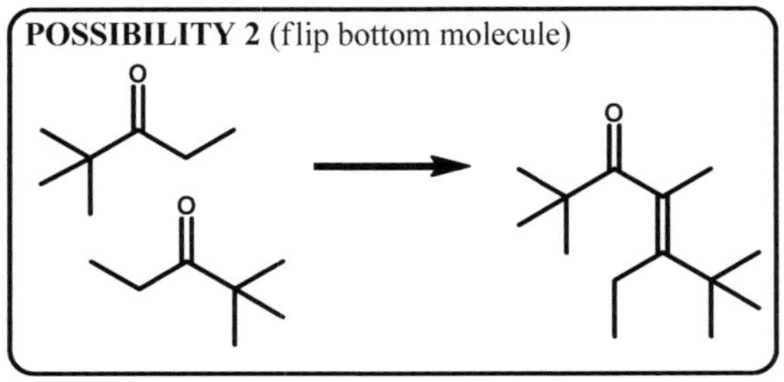

e. In Possibility 1, the product is answer choice A. Choice A has a lot of steric hindrance between the two *t*-butyl groups (indicated by the dotted line oval) compared to the product of Possibility 2, which is choice B. This makes choice B the more stable product.

The answer is B.

16. *Necessary Knowledge and Reasoning:*

a. The Michael reaction is a conjugate addition (1,4-addition) of a nucleophile to an alpha,beta-unsaturated carbonyl (α,β-unsaturated carbonyl).

b. The net product of 1,4-addition of a nucleophile to an α,β-unsaturated carbonyl features the nucleophile added to the beta position of the carbonyl and loss of the C=C bond:

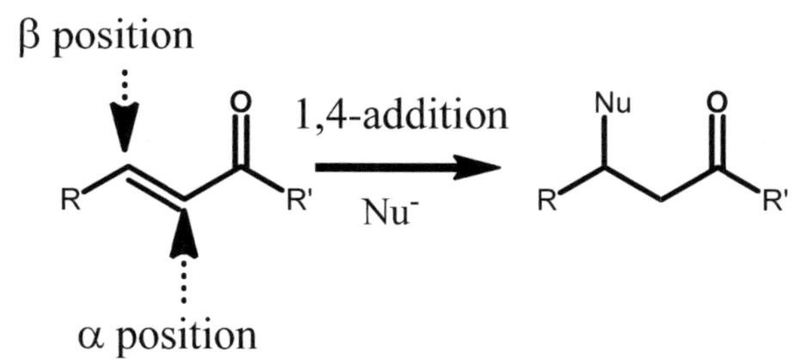

c. In this case, the nucleophile will be an enolate (an alpha-deprotonated carbonyl), so the reaction will occur like this:

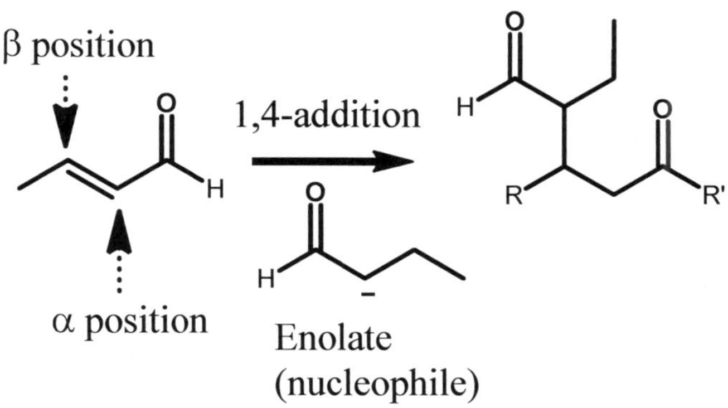

This product matches that given in answer choice D.

The answer is D.

17. *Necessary Knowledge and Reasoning:*

a. The S_N2 reaction involved nucleophilic attack at the carbon that has a leaving group on it at the same time that the leaving group leaves as an anion. The process is concerted (one single step), so there are no intermediates.

b. The nucleophile will more quickly attack a carbon that is less sterically hindered.

c. A primary site is less sterically hindered than a secondary site, which is less sterically hindered than a tertiary site.

d. Choice D has the leaving group (Br) on a primary carbon. This means that it will react faster than the substrates in choices A, B or C.

The answer is D.

18. *Necessary Knowledge and Reasoning:*

a. Water is not a very good nucleophile

b. In the absence of a good nucleophile the S_N1 reaction is favored over the S_N2 reaction.

c. The S_N1 reaction proceeds via a carbocation intermediate.

d. A carbocation has three groups coming off of the positively-charged carbon, and that site is thus achiral.

e. One cannot get a single chiral center at a site using achiral starting materials or intermediates.

f. Therefore, if the product of an S_N1 reaction has a stereogenic atom at the site that was attacked by the nucleophile, a mixture of configurations will be found at that site in the product. This is sometimes referred to as the "stereorandomization" of products produced in the S_N1 reaction.

g. In this case we expect a racemic mixture of both the *R*- and *S*-configurations.

The answer is C.

19. *Necessary Knowledge and Reasoning:*

a. When NaI is placed in solution, it will produce iodide (I^-). Iodide is a good nucleophile.

b. In the presence of a good nucleophile on a secondary substrate the S_N2 reaction is favored over the S_N1 reaction.

c. The S_N2 reaction is concerted.

d. The nucleophile attacks from the opposite side compared to where the leaving group is attached, leading to an **inversion of configuration**. This inversion of configuration is sometimes called the **Walden inversion**.

e. The starting material features the *R*-configuration about the stereogenic carbon that will be attacked by the nucleophile.

f. After inversion, we therefore expect the product to exhibit the *S*-configuration.

The answer is B.

20. *Necessary Knowledge and Reasoning:*

a. Water is not a very good nucleophile

b. The answer choices are only substitution products, so we will not need to consider elimination routes (though they will happen as well). In the absence of a good nucleophile the S_N1 reaction is favored over the S_N2 reaction.

c. The S_N1 reaction proceeds via a carbocation intermediate.

d. If the carbocation is next to a more-substituted site, the majority of it will rearrange prior to substitution.

e. In the current case, the initially-formed carbocation is secondary and is next to a quaternary carbon. A 1,2-methyl shift will occur to produce a tertiary carbocation:

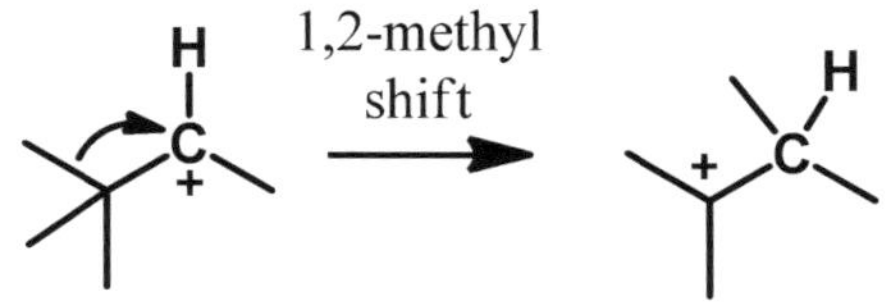

Initial Cation

f. The new tertiary carbocation will then react with water to produce an alcohol:

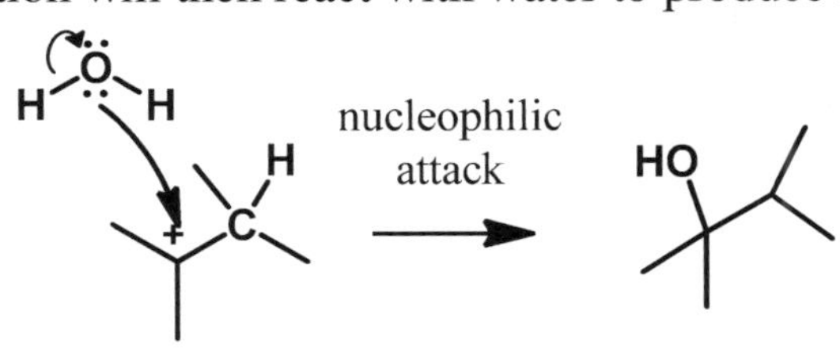

The structure of this alcohol matches that shown in choice A.

The answer is A.

21. *Necessary Knowledge and Reasoning:*

a. A Claisen condensation is simply a nucleophilic addition/elimination sequence in which an ester reacts with a nucleophile that is the enolate of another ester molecule, as shown in the next few steps.

b. In this case, there are two different esters, but only one of the two has removable protons on a carbon that is α to the carbonyl. Therefore, we will use this enolizable ester to make our enolate nucleophile:

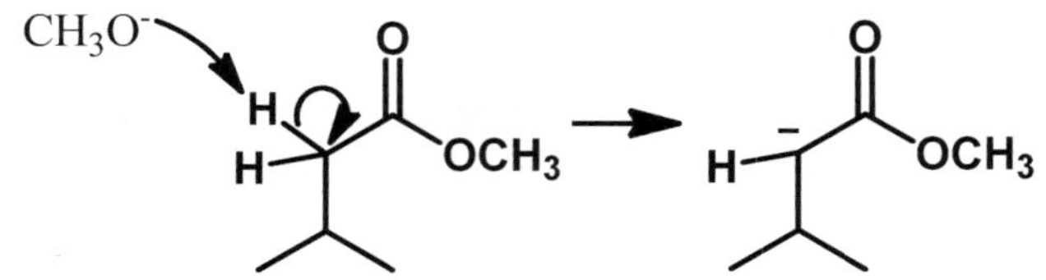

b. Now the enolate will act as the nucleophile to attack the other ester in the type of nucleophilic addition/elimination that esters generally undergo under such conditions:

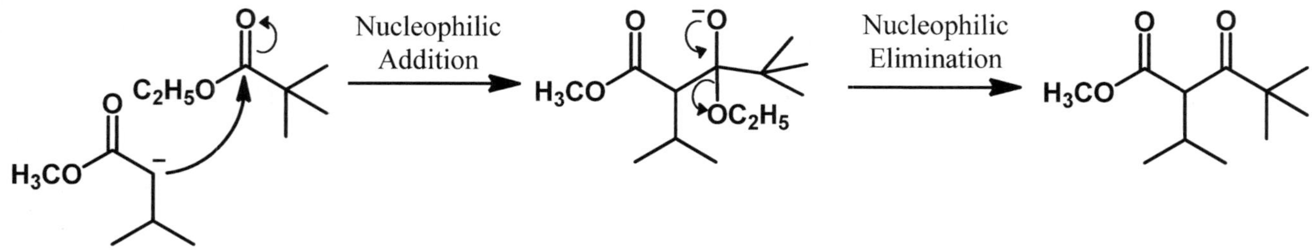

The product so formed matches the structure in choice A.

The answer is A.

22. *Necessary Knowledge and Reasoning:*

a. The starting material is an α,β-unsaturated carbonyl.

b. Gilman reagents ($LiCuR_2$) act as if they donate "R^-" groups that are nucleophiles but weak bases. Here the R group is an ethyl substituent.

c. Nucleophiles that are weak bases undergo conjugate addition (also called 1,4-addition) when they react with α,β-unsaturated carbonyls.

d. The *net* product of conjugate addition to the α,β-unsaturated carbonyl is change of the C=C bond to a single bond and placement of the nucleophile at the β-position of the α,β-unsaturated carbonyl. The answer choice in which an ethyl group has added to the β-position is choice D.

The answer is D.

23. *Necessary Knowledge and Reasoning:*

a. An *E*-alkene is one in which the higher priority substituent on each doubly-bound carbon point in opposite directions (they *anti-* to one another).

b. Prioritization is done using the Cahn-Ingold-Prelog rules. These are the same set of rules we use when prioritizing substituents in order to assign configuration to chiral molecules.

c. If one assigns all of the relevant priorities for all of the choices, only choice B features an *anti-* arrangement of the two high priority groups.

The answer is B.

24. *Necessary Knowledge and Reasoning:*

A quick way to assign hybridization to an atom is to figure out the sum of its substituents (attached atoms + lone pairs). Remember that lone pairs are not always drawn in a structure, so be sure to draw in any lone pairs needed to give the atom the formal charge shown in the structure that you are examining. Once you find the sum of its substituents (attached atoms + lone pairs), you assign hybridization as:

If (attached atoms + lone pairs) = 4, then hybridization = sp^3
If (attached atoms + lone pairs) = 3, then hybridization = sp^2
If (attached atoms + lone pairs) = 2, then hybridization = sp

The answer is D.

25. *Necessary Knowledge and Reasoning:*

a. In introductory organic chemistry courses, two routes for converting alkynes to alkenes are usually introduced: hydrogenation in the presence of Lindlar's catalyst or reaction of the alkyne with an alkali metal (Na or K) in NH_3(l).

b. One of these routes is provided in choice B.

The answer is B.

26. *Necessary Knowledge and Reasoning:*

a. The Diels-Alder reaction takes a diene and a dienophile (something with a C-C pi bond) and creates a six membered ring. If the dienophile is an alkene, the six-membered ring is a cyclohexene:

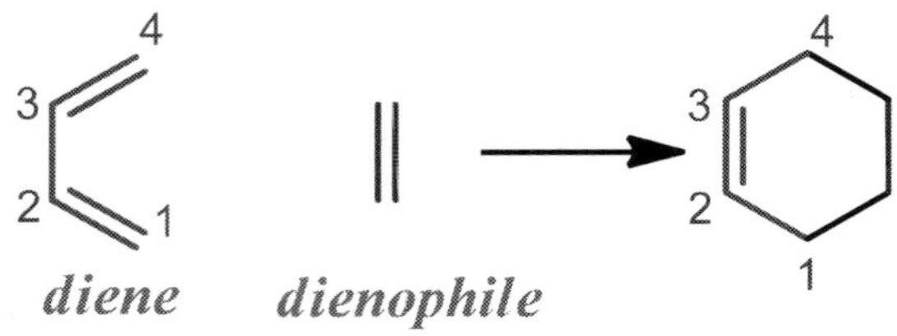

b. If there is a substituent on the diene, it will 'bend out of the way' upon reaction and attain the correct tetrahedral geometry about the sp^3 carbons (numbers 1 and 4 in the scheme below):

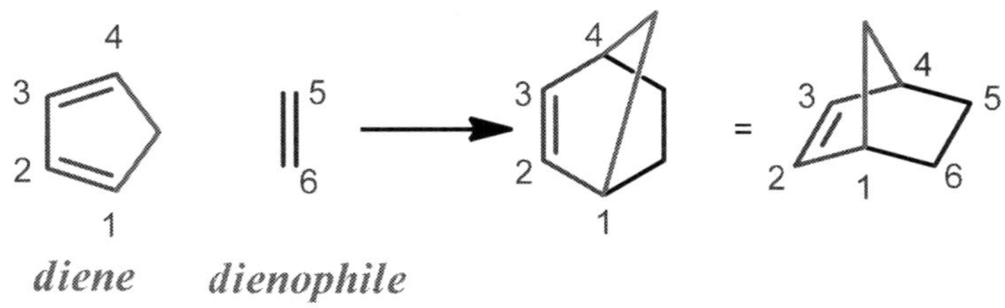

c. In the current case, there is an additional substituent, an aldehyde unit, on the product. This aldehyde unit is on C-5, which came from the dienophile, not the diene. Therefore, the diene that we need in order to create the target product is 1,3-cyclopentadiene, shown on the left above.

The answer is D.

27. *Necessary Knowledge and Reasoning:*

a. Let us consider each potential answer. Enantiomers are nonsuperimposible mirror images of one another. The two structures in question are not mirror images of each other, so they cannot be enantiomers. This rules out choice A.

b. Resonance structures involve only the movement of lone pairs and pi bonds. In this case there are actually atoms moved around. In the structure on the left, for example, there is a hydrogen atom on the oxygen, whereas this hydrogen is no longer on the oxygen in the structure on the right. This observation allows us to rule out the two being resonance structures of the same molecule. Choice B is ruled out.

c. Configurational isomers by definition have the same atom connectivity. In part b, above, we just saw that there is no way that the two compounds can be configurational isomers of one another because atom connectivity has changed. This rules out choice C.

d. Tautomers are constitutional isomers that are rapidly interconvertible via an equilibrium reaction. This specific case is an example of tautomers wherein one tautomer is an enol and the other is the keto form of that enol.

The answer is D.

28. *Necessary Knowledge and Reasoning:*

a. The best way to approach a multistep synthesis problem is to consider one step at a time. Here, the starting material is a ketone and the first step is to react it with a primary amine with catalytic acid (the abbreviation "cat." indicates a catalytic amount of something).

b. Under acid-catalyzed conditions, ketones will often lose water, which is made from the O of the C=O bond with two protons provided by the other reactant(s). A few of the general reaction types that are usually covered in an introductory organic course are:

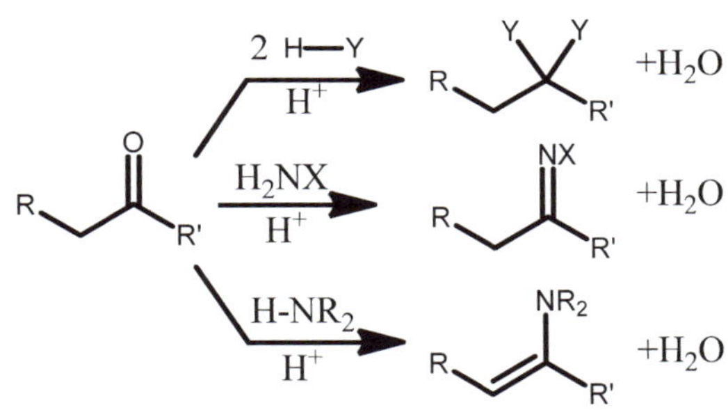

Hydrate or Ketal formation:
Example: Y = OH or OR

Double bond to Nitrogen: requires TWO protons on N in the starting material

Enamine reaction; only ONE proton on the nitrogen in the starting material

It is worth noting that aldehydes can undergo the same kind of reactions.

c. When the reactant is a ketone or an aldehyde, one should consider all of these possibilities. In the current case, our first reactant features a nitrogen with TWO protons on it. This will allow for a net reaction like the second pathway shown above. We would thus expect the product to feature a double bond between nitrogen and the formerly-carbonyl carbon:

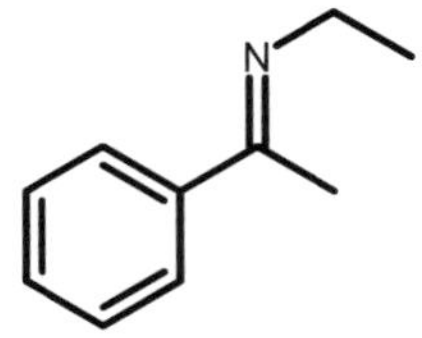

This imine (also called a Schiff base) will be the product of our first step and will serve as the starting material for step 2.

d. Imines react very similarly to carbonyls in many cases.

e. Our second step features $LiAlH_4$ as a reactant. This species will produce hydride anions (H^-) as the putative reactive species. Hydride is a very good nucleophile and will react with a polar pi bond by nucleophilic addition like this:

CARBONYL:

IMINE:

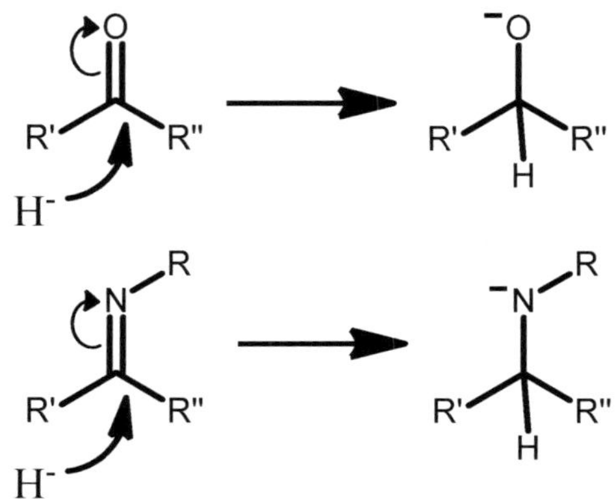

f. We would thus expect that the product of our second step would be:

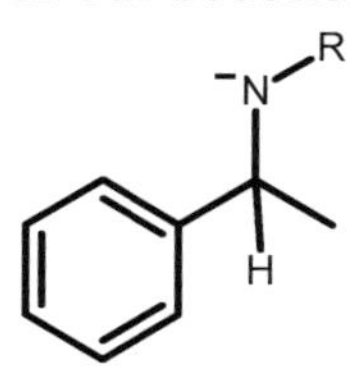

g. In the final step H^+ will simply attack the anionic nitrogen to make the neutral amine. This produces the compound shown in choice C (note that Ph stands for phenyl, the name given to benzene when it is a substituent).

The answer is C.

29. *Necessary Knowledge and Reasoning:*

a. A Lewis Base is a lone pair donor.

b. In order for a nitrogen to be effective at donating its lone pair, the lone pair cannot be tied up in resonance or in an aromatic system. One can draw resonance structures for choice C (pyrrole) in which the lone pair on the N is donated into the pi system to make it aromatic. This makes pyrrole a poor base.

c. If what looks like an sp^3 hybridized atom with a lone pair is adjacent to an atom that is engaged in a pi bond (a double or triple bond), then the sp^3 lone pair will be tied up in a resonance structure. A common example is with nitrogen, such as in choice A (aniline). In aniline, nitrogen's sp^3 lone pair will be involved in resonance with the nearby phenyl ring, thereby making the lone pair less available to bond with H^+. Consequently, nitrogens such as this are not good bases.

d. If an atom with a lone pair is sp^2 hybridized, it cannot donate its lone pair to an adjacent pi system, so the lone pair is available for donation to an acid. Choice B (pyridine) features an sp^2-hybridized nitrogen that could possibly donate its lone pair. Pyridine is a moderately good base for organic reactions and cannot be ruled out yet.

e. A nucleus has a greater attraction for electrons in an sp^2 orbital than in an sp^3 orbital, so it is easier for an sp^3 hybridized atom to donate its electrons (which is necessary for it to act as a base). This fact leads us to realize that the sp^2 orbital of pyridine (choice B) will hold onto its lone pair more strongly than the sp^3-hybridized orbital will hold onto the lone pair on the N in choice D. For this reason, choice D will more easily donate its lone pair and it is thus the stronger base.

The answer is D.

30. *Necessary Knowledge and Reasoning:*

a. This reaction features a very unusual reagent, a phosphorus ylide, reacting with an aldehyde.

b. In introductory organic chemistry courses, the phosphorus ylide is generally only used in the Wittig reaction, which produces an alkene. In this reaction, the carbon with the formal charge of -1 will end up doubly bound to the carbonyl carbon in place of the oxygen on the carbonyl. The phosphorus-containing piece ends up doubly bound to the oxygen instead:

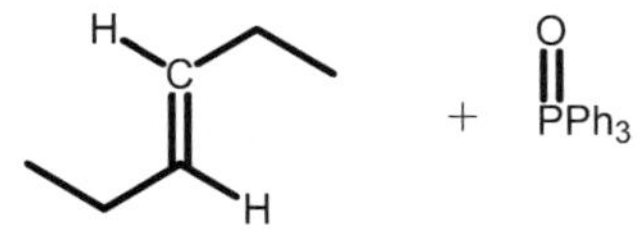

Note that the correct major product will be the *E*-alkene rather than the *Z*-alkene because there is less steric hindrance in the *E*-alkene. The mechanism is several steps and should be reviewed in your reaction book to understand this process more thoroughly if desired.

The answer is B.

31. *Necessary Knowledge and Reasoning:*

a. The best way to approach a multistep synthesis problem is to consider one step at a time. The first reaction is between $SOCl_2$ (thionyl chloride) and a carboxylic acid. This is a very good way to exchange the poor leaving group –OH of the carboxylic acid for a good leaving group –Cl. This is often referred to as an "activation" reaction because it turns the carboxylic acid into a much more reactive acid chloride. The product of this first step is thus:

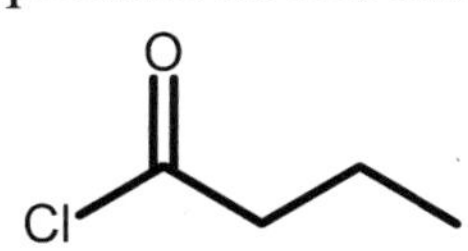

Note that this is choice B, but it is not the product of the whole series of reactions, so we cannot choose it until we work through all the other steps.

b. Acid chlorides react readily with a wide variety of nucleophiles via a nucleophilic addition-nucleophilic elimination sequence wherein the Cl is replaced by the nucleophile by this general reaction:

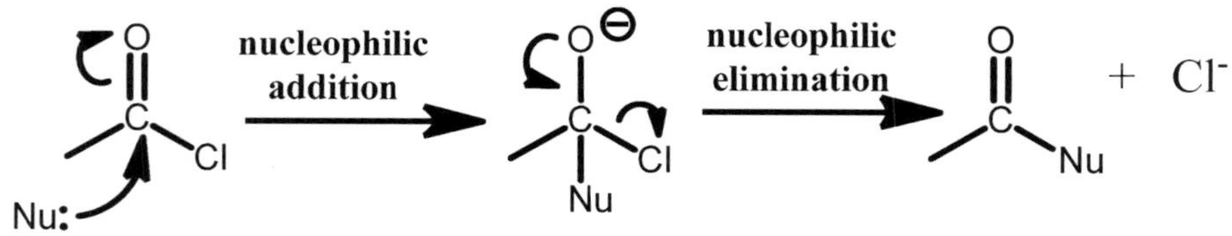

c. In this case, it is the $–OCH_3$ portion of the methanol reactant that will replace the Cl. This means that the product of the second step is:

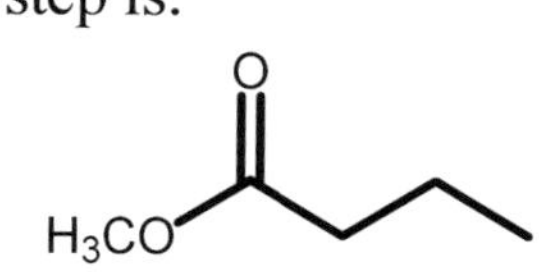

d. The next reaction employs DIBALH as the reactant. DIBALH is an abbreviation for diidobutylaluminum hydride. DIBALH is selective for adding ONE equivalent of hydride to a carbonyl. In this ester, the $-OCH_3$ leaving group ends up replaced by a hydrogen from the DIBALH. This is an example of the classic use of DIBALH to make an aldehyde:

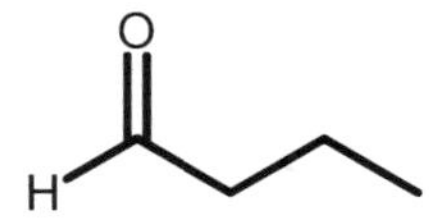

This matches the structure provided in choice D.

The answer is D.

32. *Necessary Knowledge and Reasoning:*

a. Decarboxylation occurs when you have one carbonyl carbon that has one carbon between it and the carbon of a carboxylic acid:

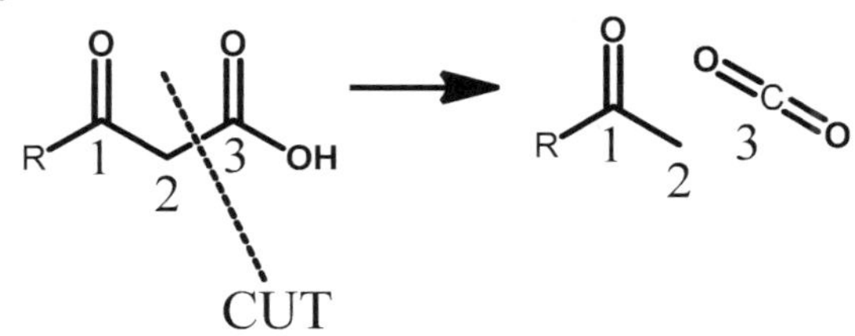

b. Choices B and C do not have carboxylic acids in them, so we rule those out.

c. Choice D has a carboxylic acid in it, but it is separated from the other carbonyl by *two* carbons, so we rule it out. This leaves us with choice A as the best selection.

The answer is A.

33. *Necessary Knowledge and Reasoning:*

a. As the question states, deuterium acts like H in the reactions that we have learned. This means that LiAl**D**$_4$ will act like LiAl**H**$_4$.

b. $LiAlH_4$ produces hydride anions (H^-) as the putative reactive species. Likewise, $LiAlH_4$ produces **deuteride** anions (D^-) as the putative reactive species. Deuteride, like hydride, is a very good nucleophile, and will react with a polar pi bond by nucleophilic addition like this:

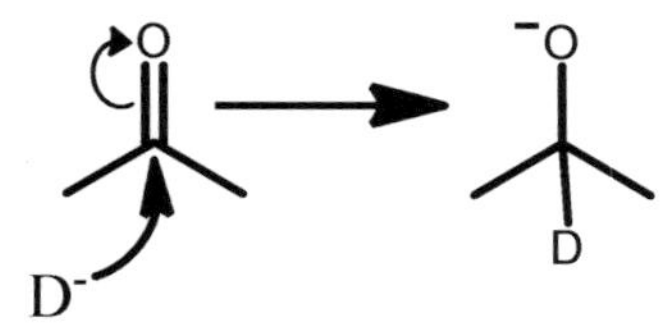

b. The next step is to add H^+. This is done to put a proton onto the anionic oxygen from the first step to produce the final neutral product:

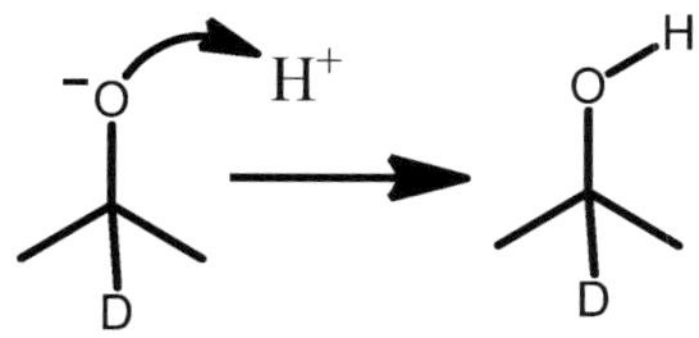

The answer is C.

34. *Necessary Knowledge and Reasoning:*

a. The first reaction conditions, $NaNO_2$ with an acid, are characteristic of a **diazotization reaction**.

b. A diazotization reaction first converts the NH_2 unit of an aromatic amine into a diazonium salt:

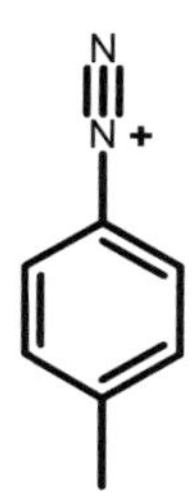

b. The diazonium unit is readily displaced by a number of anions because N_2 gas is produced in a very favorable reaction upon displacement.

c. In this case, CuBr is added after the diazonium salt is formed. The anionic portion of this reagent is bromide, so our final product will feature net replacement of NH_2 with Br. Choice C provides this option.

The answer is C.

35. *Necessary Knowledge and Reasoning:*

a. The general Malonic ester synthesis is:

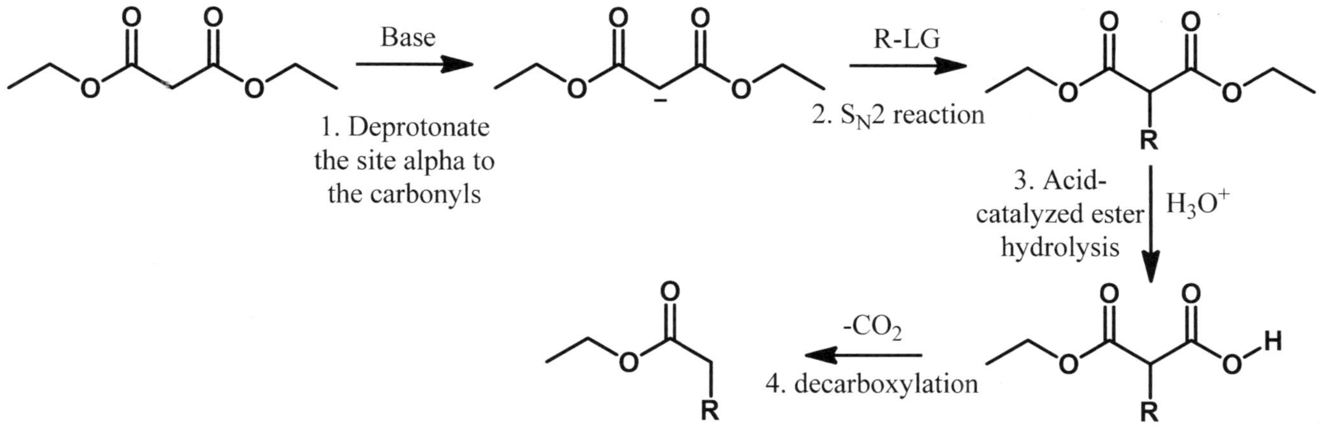

Note that this reaction is actually a stringing together of several simpler reactions. This specific sequence has a special name because it has proven so important industrially. LG is an abbreviation for "leaving group".

b. From the general reaction, we see that the starting material and the conditions are all constant other than the identity of the R group with the leaving group on it. We see from the above scheme that the final product features an $-OCH_2CH_3$ unit on one side of the carbonyl and a chain that is the R group plus one carbon on the other side:

General Product:

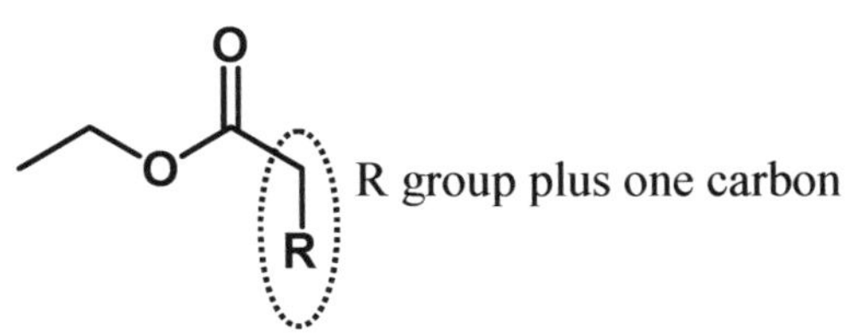

Specific Product in this Problem:

c. If the "R group plus one carbon"-long chain is four carbons in length, then the R group we must use to get this product is three carbons long (a propyl group). This points us to choose answer C.

The answer is C.

28459860R00151

Made in the USA
San Bernardino, CA
30 December 2015